EVERYDAY ENVIRONMENTAL TOXINS
TOXINS

Children's Exposure Risks

EVERYDAY ENVIRONMENTAL TOXINS

Children's Exposure Risks

Edited by
Areej Hassan, MD, MPH

Apple Academic Press Inc.	Apple Academic Press Inc.
3333 Mistwell Crescent	9 Spinnaker Way
Oakville, ON L6L 0A2	Waretown, NJ 08758
Canada	USA

© 2015 by Apple Academic Press, Inc.

First issued in paperback 2021

Exclusive worldwide distribution by CRC Press, a member of Taylor & Francis Group

No claim to original U.S. Government works

ISBN 13: 978-1-77463-375-5 (pbk)
ISBN 13: 978-1-77188-101-2 (hbk)

Library of Congress Control Number: 2014952125

Library and Archives Canada Cataloguing in Publication

Everyday environmental toxins : children's exposure risks/edited by Areej Hassan, MD, MPH.

Includes bibliographical references and index.
ISBN 978-1-77188-101-2 (bound)
1. Pediatric toxicology. 2. Environmental toxicology. 3. Children--Health risk assessment. 4. Children and the environment. I. Hassan, Areej, editor

RA1225.E94 2015   615.90083   C2014-906646-5

Apple Academic Press also publishes its books in a variety of electronic formats. Some content that appears in print may not be available in electronic format. For information about Apple Academic Press products, visit our website at **www.appleacademicpress.com** and the CRC Press website at **www.crcpress.com**

ABOUT THE EDITOR

AREEJ HASSAN, MD, MPH

Areej Hassan MD, MPH, is an attending physician at Boston Children's Hospital, Bostor, Massachusetts, USA. She completed her pediatric residency at Hasbro Children's Hospital, Providence, Rhode Island prior to training in adolescent medicine at Boston Children's. In addition to primary care, she focuses her clinical interests on reproductive endocrinology and international health. She also maintains an active role in medical education and has particular interest in building and developing innovative teaching tools through open educational resources. She currently teaches, consults, and is involved in pediatric and adolescent curricula development at multiple sites abroad in Central America and Southeast Asia.

CONTENTS

ACKNOWLEDGMENT AND HOW TO CITE

The editor and publisher thank each of the authors who contributed to this book, whether by granting their permission individually or by releasing their research as open source articles or under a license that permits free use, provided that attribution is made. The chapters in this book were previously published in various places in various formats. To cite the work contained in this book and to view the individual permissions, please refer to the citation at the beginning of each chapter. Each chapter was read individually and carefully selected by the editor; the result is a book that provides a nuanced study of environmental toxins and the risks they pose to children's development. The chapters included examine the following topics:

- Chapters 1 and 2 lay the groundwork for this book, explaining why exposures during the childhood period are so critical to address and prevent. The authors tackle an overwhelming topic by breaking it down into categories and developmental milestones. The tables and figures offer as much clear information as the text. Although the focus is very heavy on the conceptional, prenatal windows, the authors explain that this limitation is caused by limited data from the pre-pubertal and pubertal years.
- Chapter 3, a well-written editorial, provides interesting background behind the development of this field. It includes valuable references to specific tools and teaching materials available for clinicians.
- Epigenetics is a growing field of investigation, making this article particularly timely, even if it has less practical application for the clinician. The authors' research in Chapter 4 indicates a significant association between prenatal smoke exposure and genetic changes.
- The well-designed and well-executed study in Chapter 5 offers well-defined outcome measures pertaining to the opposite end of the research spectrum from the previous article. The research here focuses on the immediate effects of tobacco smoke on children's health.
- Chapter 6 is another well-written and well-designed study. Again, the results are surprising, but even negative outcomes are important findings.
- Chapter 7 provides a well-written discussion with striking results, while also addressing limitations. This is research that should be used to inform

future policy regarding housing, urban development, clean energy, and a myriad other issues.

- Chapter 8 provides a review of recent studies that suggests numerous environmental risk factors could be critical for lead exposure, information that is vital to clinicians and policymakers.
- Chapter 9 is an important investigation of the ways in which even low levels of lead exposure can impact health.
- Chapter 10 article lays the groundwork for the far more thorough study that follows next.
- Chapter 11 provides an example of beautifully done research with long-term outcomes linked to exposure measured through blood. Studies like this one should be released to the media and used to inform policy.
- The two articles in this section, Chapters 12 and 13, were chosen to indicate a natural progression, starting first with prenatal exposure and then looking at adolescent levels of exposure. Results are consistent along the spectrum: the higher the exposure as indicated by blood levels, the greater the impact of neurodevelopmental outcomes.
- In Chapter 14, we move from studies looking at the impact of toxins in real time to looking at the incredibly disturbing fact that toxins can impact generations to come.
- With Chapter 15, the book concludes with an article by the World Health Organization, which offers a practical and comprehensive summary of a series of action steps.

LIST OF CONTRIBUTORS

John L. Adgate
Department of Environmental and Occupational Health

Cem Akkus
Department of Earth Sciences, University of Memphis, Memphis, TN 38152, USA

Srikesh Arunajadai
Department of Biostatistics, Mailman School of Public Health

Willy Baeyens
Department of Analytical and Environmental Chemistry, Brussels Free University (VUB), Brussels, Belgium

Kathleen C. Barnes
Johns Hopkins University, Baltimore, Maryland, United States of America

Dana B. Barr
Emory University, Atlanta, Georgia, USA

S. Boese-O'Reilly
Children's Environmental Health Professional, Munich, Germany

Carrie V. Breton
Department of Preventive Medicine, Keck School of Medicine, University of Southern California, Los Angeles, California, United States of America

John W. Brock
National Center for Environmental Health, Centers for Disease Control and Prevention, Atlanta, Georgia, USA

Liesbeth Bruckers
Interuniversity Institute for Biostatistics and Statistical Bioinformatics, Hasselt University, Diepenbeek, Belgium

Vincent Carey
Channing Division of Network Medicine, Department of Medicine, Brigham and Women's Hospital, Harvard Medical School, Boston, Massachusetts, United States of America

Adrian Covaci
Toxicological Center, University of Antwerp, Antwerp, Belgium

Kim Croes
Department of Analytical and Environmental Chemistry, Brussels Free University (VUB), Brussels, Belgium

Lea A. Cupul-Uicab
Epidemiology Branch, National Institute of Environmental Health Sciences, National Institutes of Health, Department of Health and Human Services, Research Triangle Park, North Carolina, USA and Center for Population Health Research, National Institute of Public Health, Cuernavaca, Morelos, Mexico

Johan C. de Jongste
Department of Paediatrics, Division of Respiratory Medicine, Erasmus Medical Center-Sophia Children's Hospital, Rotterdam, The Netherlands and Department of General Practice and Elderly Care Medicine, EMGO, VU University Medical Center, Amsterdam, The Netherlands

Harry J. de Koning
Department of Public Health, Erasmus Medical Center, Rotterdam, The Netherlands

Lora D. Delwiche
Department of Public Health Sciences, University of California, Davis, Davis, California, USA

Elly Den Hond
Flemish Institute for Technological Research, Environmental Risk and Health, Mol, Belgium

Alin C. Dirtu
Toxicological Center, University of Antwerp, Antwerp, Belgium

Liesbeth Duijts
Department of Paediatrics, Division of Respiratory Medicine, Erasmus Medical Center-Sophia Children's Hospital, Rotterdam, The Netherlands, Department of Epidemiology, Erasmus Medical Center, Rotterdam, The Netherlands and Department of Paediatrics, Division of Neonatology, Erasmus Medical Center-Sophia Children's Hospital, Rotterdam, The Netherlands

Ruth A. Etzel
Senior Officer for Environmental Health Research, Department of Public Health and Environment, World Health Organization, Geneva, Switzerland

Gabriel M. Filippelli
Department of Earth Sciences, Department of Public Health, and Center for Urban Health, Indiana University-Purdue University, Indianapolis (IUPUI), Indianapolis, Indiana 46202, United States

A. Garg
Center for Children's Health and the Environment, Department of Community and Preventative Medicine, Mount Sinai School of Medicine, New York, USA

Estella M. Geraghty
Division of General Medicine, School of Medicine, University of California, Davis, Sacramento, California, USA

Frank Gilliland
Department of Preventive Medicine, Keck School of Medicine, University of Southern California, Los Angeles, California, United States of America

Ruixin Guo
Department of Biostatistics and Informatics, Colorado School of Public Health, Aurora, Colorado, USA

Siri E. Håberg
Norwegian Institute of Public Health, Oslo, Norway

Esther Hafkamp-de Groen
The Generation R Study Group, Erasmus Medical Center, Rotterdam, The Netherlands and Department of Public Health, Erasmus Medical Center, Rotterdam, The Netherlands

Robin L. Hansen
Department of Pediatrics, School of Medicine, University of California, Davis, Sacramento, California, USA and UC Davis Medical Investigations of Neurodevelopmental Disorders (MIND) Institute, Sacramento, California, USA

Julie B. Herbstman
Columbia Center for Children's Environmental Health, Department of Environmental Health Sciences, Mailman School of Public Health, Columbia University, New York, New York, USA

Irva Hertz-Picciotto
Department of Public Health Sciences, University of California, Davis, Davis, California, USA and UC Davis Medical Investigations of Neurodevelopmental Disorders (MIND) Institute, Sacramento, California, USA

Lori Hoepner
Columbia Center for Children's Environmental Health, Mailman School of Public Health, Columbia University, New York, New York, USA

Albert Hofman
Department of Epidemiology, Erasmus Medical Center, Rotterdam, The Netherlands

Megan Horton
Sergievsky Center and Columbia Center for Children's Environmental Health, Mailman School of Public Health, Columbia University, New York, New York, USA

Vincent W. Jaddoe
The Generation R Study Group, Erasmus Medical Center, Rotterdam, The Netherlands, Department of Epidemiology, Erasmus Medical Center, Rotterdam, The Netherlands, and Department of Paediatrics, Erasmus Medical Center-Sophia Children's Hospital, Rotterdam, The Netherlands

Richard S. Jones
Division of Laboratory Sciences, National Center for Environmental Health, Centers for Disease Control and Prevention, Atlanta, Georgia, USA

Bonnie R. Joubert
Division of Intramural Research, National Institute of Environmental Health Sciences, National Institutes of Health, Dept of Health and Human Services, Research Triangle Park, North Carolina, United States of America

Michał Kiciński
Centre for Environmental Sciences, Hasselt University, Diepenbeek, Belgium

C. A. Kimmel
National Center for Environmental Assessment/ORD, Environmental Protection Agency, Washington, NC, USA

Mark A. Klebanoff
The Ohio State University College of Medicine, Columbus, Ohio, USA and The Research Institute at Nationwide Children's Hospital, Columbus, Ohio, USA

Matthew Kurzon
Columbia Center for Children's Environmental Health, Department of Environmental Health Sciences, Mailman School of Public Health, Columbia University, New York, New York, USA

Mark A. S. Laidlaw
Environmental Science, Department of Environment and Geography, Faculty of Science, Macquarie University, Sydney, New South Wales, 2109, Australia

P. J. Landrigan
Center for Children's Health and the Environment, Department of Community and Preventative Medicine, Mount Sinai School of Medicine, New York, USA

Sally A. Lederman
Columbia Center for Children's Environmental Health, Department of Environmental Health Sciences, Mailman School of Public Health, Columbia University, New York, New York, USA

Robert Lemanske
University of Wisconsin, Madison, Wisconsin, United States of America

Andy Liu
National Jewish Health, Denver, Colorado, United States of America

Stephanie London
Division of Intramural Research, National Institute of Environmental Health Sciences, National Institutes of Health, Dept of Health and Human Services, Research Triangle Park, North Carolina, United States of America

Matthew P. Longnecker
Epidemiology Branch, National Institute of Environmental Health Sciences, National Institutes of Health, Department of Health and Human Services, Research Triangle Park, North Carolina, USA

Fernando Martinez
Arizona Respiratory Center, University of Arizona, Arizona, United States of America

Shawn P. McElmurry
Department of Civil and Environmental Engineering, Wayne State University, Detroit, Michigan 48202, United States

Lisa M. McKenzie
Department of Environmental and Occupational Health

P. Mendola
National Health and Environmental Effects Research Laboratory/ORD, Environmental Protection Agency, Research Triangle Park, NC, USA

Ashna D. Mohangoo
TNO, Netherlands Organisation for Applied Scientific Research, Department of Child Health, Leiden, The Netherlands

Tim S. Nawrot
Centre for Environmental Sciences, Hasselt University, Diepenbeek, Belgium and School of Public Health, Occupational and Environmental Medicine, KULeuven, Leuven, Belgium

Larry L. Needham
Division of Laboratory Sciences, National Center for Environmental Health, Centers for Disease Control and Prevention, Atlanta, Georgia, USA

Vera Nelen
Department of Health, Provincial Institute for Hygiene, Antwerp, Belgium

Lee S. Newman
Department of Environmental and Occupational Health

Dan Nicolae
University of Chicago, Chicago, Illinois, United States of America

Megan Niedzwiecki
Columbia Center for Children's Environmental Health, Department of Environmental Health Sciences, Mailman School of Public Health, Columbia University, New York, New York, USA

Wenche Nystad
Norwegian Institute of Public Health, Oslo, Norway

Carole Ober
University of Chicago, Chicago, Illinois, United States of America

Esra Ozdenerol
Department of Earth Sciences, University of Memphis, Memphis, TN 38152, USA

Frederica Perera
Columbia Center for Children's Environmental Health, Mailman School of Public Health, Columbia University, New York, New York, USA

Weiliang Qui
Channing Division of Network Medicine, Department of Medicine, Brigham and Women's Hospital, Harvard Medical School, Boston, Massachusetts, United States of America

Hein Raat
Department of Public Health, Erasmus Medical Center, Rotterdam, The Netherlands

Virginia Rauh
Columbia Center for Children's Environmental Health, Department of Environmental Health Sciences, Mailman School of Public Health, Columbia University, New York, New York, USA and Heilbrunn Center for Population and Family Health, Mailman School of Public Health

Beate Ritz
Departments of Epidemiology, Environmental Health Sciences and Neurology, Fielding School of Public Health and School of Medicine, University of California, Los Angeles, Los Angeles, California, USA

David A. Savitz
Department of Epidemiology, Brown University, Providence, Rhode Island, USA

Charles W. Schmidt
An award-winning science writer from Portland, ME who has written for Discover Magazine, Science, and Nature Medicine.

Rebecca J. Schmidt
Department of Public Health Sciences, University of California, Davis, Davis, California, USA

Greet Schoeters
Flemish Institute for Technological Research, Environmental Risk and Health, Mol, Belgium and Department of Biomedical sciences, University of Antwerp, Antwerp, Belgium

S. G. Selevan
National Center for Environmental Assessment/ORD, Environmental Protection Agency, Washington, NC, USA

Janie F. Shelton
Department of Public Health Sciences, University of California, Davis, Davis, California, USA

M. K. E. Shimkin
MShimkin Consulting, Alexandria, USA

Kimberly D. Siegmund
Department of Preventive Medicine, Keck School of Medicine, University of Southern California, Los Angeles, California, United States of America

Isabelle Sioen
Department of Public Health, Ghent University, Ghent, Belgium

Andreas Sjödin
Division of Laboratory Sciences, National Center for Environmental Health, Centers for Disease Control and Prevention, Atlanta, Georgia, USA

Robert Strunk
Washington University School of Medicine, St. Louis, Montana, United States of America

Daniel J. Tancredi
Department of Pediatrics, School of Medicine, University of California, Davis, Sacramento, California, USA and Center for Healthcare Policy and Research, School of Medicine, University of California, Davis, Sacramento, California, USA

Deliang Tang
Columbia Center for Children's Environmental Health, Department of Environmental Health Sciences, Mailman School of Public Health, Columbia University, New York, New York, USA

Mark Taylor
Environmental Science, Department of Environment and Geography, Faculty of Science, Macquarie University, Sydney, New South Wales, 2109, Australia

Ralf J. P. van der Valk
The Generation R Study Group, Erasmus Medical Center, Rotterdam, The Netherlands, Department of Paediatrics, Division of Respiratory Medicine, Erasmus Medical Center-Sophia Children's Hospital, Rotterdam, The Netherlands, and Department of Epidemiology, Erasmus Medical Center, Rotterdam, The Netherlands

Nicolas Van Larebeke
Department of Radiotherapy and Nuclear Medicine, University Ghent, Ghent, Belgium

Mineke K. Viaene
Department of Neurology, Sint Dimphna Hospital, Geel, Belgium

Richard Y. Wang
Division of Laboratory Sciences, National Center for Environmental Health, Centers for Disease Control and Prevention, Atlanta, Georgia, USA

Xinhui Wang
Department of Preventive Medicine, Keck School of Medicine, University of Southern California, Los Angeles, California, United States of America

Scott Weiss
Channing Division of Network Medicine, Department of Medicine, Brigham and Women's Hospital, Harvard Medical School, Boston, Massachusetts, United States of America

Robin Whyatt
Columbia Center for Children's Environmental Health, Mailman School of Public Health, Columbia University, New York, New York, USA

Roxana Z. Witter
Department of Environmental and Occupational Health

Sammy Zahran
Department of Economics, Colorado State University, Fort Collins, Colorado 80523, United States and Robert Wood Johnson Health and Society Scholar, Department of Epidemiology, Columbia University, New York, New York 10032, United States

INTRODUCTION

There is growing recognition that the impact of the environment on children's health is of critical importance for both current and future generations. In the last half-century, thousands of chemicals have been introduced into the environment with limited—although growing—research on the consequences of exposure. It is clear that children and adolescents are far more vulnerable than adults to these environmental toxins by virtue of children's behaviors, higher metabolic rate, greater skin area relative to their volume, and still developing organ systems. Increased number of ear infections, poor asthma control, and learning disabilities are just some of the adverse outcomes that have been noted.

This collection of timely articles details the impact of a number of commonplace environmental toxins, including tobacco smoke, lead, pesticides, and flame retardants, on specific health outcomes, before ending with a call to action. The juxtaposition of research results collected here will, we hope, help to create greater awareness, which in turn will not only spur continued research into this vital area of study, but will also be of use to the clinicians who must treat the effects of environmental toxins on their young patients. Perhaps most critically of all, however, the united and concentrated message of this volume as a whole should form a powerful message to government officials, putting pressure on them to develop policies that improve the quality of the environment and spare children the detrimental effects of such exposures.

Areej Hassan

In Chapter 1, an extract from a World Health Organization report titled *Children's Health and the Environment: A Global Perspective,* Landrigan and Garg describe the growing concern over environmental hazards worldwide. The article provides an overall introduction for the book, describing why children's environmental health is so important and arguing that children are much more vulnerable to environmental toxins than adults.

Chapter 2 is taken from the same WHO report as Chapter 1. In this article, Seleven and colleagues describe the periods of vulnerability to toxins throughout children's development, and argue for a "child-centered" approach to research and public policy.

Chapter 3, by Etzel, is an editorial that argues new approaches to medical care are necessary that take into account children's unique physiology and development and the increased burden of environmental risk on them. She describes the growth of the children's environmental health (CEH) movement, laying out its early origins as well as calling for future action.

Smoking while pregnant is associated with a myriad of negative health outcomes in the child. Some of the detrimental effects may be due to epigenetic modifications, although few studies have investigated this hypothesis in detail. The goal of Chapter 4, by Breton and colleagues, is to characterize site-specific epigenetic modifications conferred by prenatal smoking exposure within asthmatic children. Using Illumina HumanMethylation27 microarrays, the authors estimated the degree of methylation at 27,578 distinct DNA sequences located primarily in gene promoters using whole blood DNA samples from the Childhood Asthma Management Program (CAMP) subset of Asthma BRIDGE childhood asthmatics (n = 527) ages 5–12 with prenatal smoking exposure data available. Using beta-regression, the authors screened loci for differential methylation related to prenatal smoke exposure, adjusting for gender, age and clinical site, and accounting for multiple comparisons by FDR. Of 27,578 loci evaluated, 22,131 (80%) passed quality control assessment and were analyzed. Sixty-five children (12%) had a history of prenatal smoke exposure. At an FDR of 0.05, we identified 19 CpG loci significantly associated with prenatal smoke, of which two replicated in two independent populations. Exposure was associated with a 2% increase in mean CpG methylation in FRMD4A (p = 0.01) and C11orf52 (p = 0.001) compared to no exposure. Four additional genes, XPNPEP1, PPEF2, SMPD3 and CRYGN, were nominally associated in at least one replication group. These data suggest that prenatal exposure to tobacco smoke is associated with reproducible epigenetic changes that persist well into childhood. However, the biological significance of these altered loci remains unknown.

In Chapter 5, Hafkamp-de Groen and colleagues aimed to evaluate the effectiveness of systematic assessment of asthma-like symptoms and environmental tobacco smoke (ETS) exposure during regular preventive well-child visits between age 1 and 4 years by well-child professionals. Sixteen well-child centres in Rotterdam, the Netherlands, were randomised into 8 centres where the brief assessment form regarding asthma-like symptoms and ETS exposure was used and 8 centres that applied usual care. 3596 and 4179 children (born between April 2002 and January 2006) and their parents visited the intervention and control centres, respectively. At child's age 6 years, physician-diagnosed asthma ever, wheezing, fractional exhaled nitric oxide (FeNO), airway resistance (Rint), health-related quality of life (HRQOL) and ETS exposure at home ever were measured. Linear mixed models were applied. No differences in asthma, wheezing, FeNO, Rint or HRQOL measurements between intervention and control group were found using multilevel regression in an intention-to-treat analysis (p>0.05). Children of whom the parents were interviewed by using the brief assessment form at the intervention well-child centres had a decreased risk on ETS exposure at home ever, compared to children who visited the control well-child centres, in an explorative per-protocol analysis (aOR = 0.71, 95% CI:0.59–0.87). Systematic assessment and counselling of asthma-like symptoms and ETS exposure in early childhood by well-child care professionals using a brief assessment form was not effective in reducing the prevalence of physician-diagnosed asthma ever and wheezing, and did not improve FeNO, Rint or HRQOL at age 6 years. The results hold some promise for interviewing parents and using information leaflets at well-child centres to reduce ETS exposure at home in preschool children.

In some previous studies, prenatal exposure to persistent organochlorines such as 1,1-dichloro-2,2-bis(p-chlorophenyl)ethylene (p,p′-DDE), polychlorinated biphenyls (PCBs), and hexachlorobenzene (HCB) has been associated with higher body mass index (BMI) in children. Cupul-Uicab and colleague's goal in Chapter 6 was to evaluate the association of maternal serum levels of β-hexachlorocyclohexane (β-HCH), p,p′-DDE, dichlorodiphenyltrichloroethane (p,p′-DDT), dieldrin, heptachlor epoxide, HCB, trans-nonachlor, oxychlordane, and PCBs with offspring obesity during childhood. The analysis was based on a subsample of 1,915 children followed until 7 years of age as part of the U.S. Collaborative Perinatal Project (CPP). The CPP enrolled pregnant women in 1959–1965; exposure levels were measured in third-trimester maternal serum that was collected before these organochlorines were banned in the United States. Childhood overweight and obesity were defined using age- and sex-specific cut points for BMI as recommended by the International Obesity Task Force. Adjusted results did not show clear evidence for an association between organochlorine exposure and obesity; however, a suggestive finding emerged for dieldrin. Compared with those in the lowest quintile (dieldrin, < 0.57 µg/L), odds of obesity were 3.6 (95% CI: 1.3, 10.5) for the fourth and 2.3 (95% CI: 0.8, 7.1) for the highest quintile. Overweight and BMI were unrelated to organochlorine exposure. In this population with relatively high levels of exposure to organochlorines, no clear associations with obesity or BMI emerged.

Birth defects are a leading cause of neonatal mortality. Natural gas development (NGD) emits several potential teratogens, and U.S. production of natural gas is expanding. In Chapter 7, McKenziee and colleagues examined associations between maternal residential proximity to NGD and birth outcomes in a retrospective cohort study of 124,842 births between 1996 and 2009 in rural Colorado. The authors calculated inverse distance weighted natural gas well counts within a 10-mile radius of maternal residence to estimate maternal exposure to NGD. Logistic regression, adjusted for maternal and infant covariates, was used to estimate associations with exposure tertiles for congenital heart defects (CHDs), neural tube defects (NTDs), oral clefts, preterm birth, and term low birth weight. The association with term birth weight was investigated using multiple linear regression. Prevalence of CHDs increased with exposure tertile, with an odds ratio (OR) of 1.3 for the highest tertile (95% CI: 1.2, 1.5); NTD prevalence was associated with the highest tertile of exposure (OR = 2.0; 95% CI: 1.0, 3.9, based on 59 cases), compared with the absence of any gas wells within a 10-mile radius. Exposure was negatively associated with preterm birth and positively associated with fetal growth, although the magnitude of association was small. No association was found between exposure and oral clefts. In this large cohort, the authors observed an association between density and proximity of natural gas wells within a 10-mile radius of maternal residence and prevalence of CHDs and possibly NTDs. Greater specificity in exposure estimates is needed to further explore these associations.

Childhood exposure to lead remains a critical health control problem in the US. Integration of Geographic Information Systems (GIS) into childhood lead exposure studies significantly enhanced identifying lead hazards in the environment and determining at risk children. Research indicates that the toxic threshold for lead exposure was updated three times in the last four decades: 60 to 30 micrograms per deciliter (μg/dL) in 1975, 25 μg/dL in 1985, and 10 μb/dL in 1991. These changes revealed the extent of lead poisoning. By 2012 it was evident that no safe blood lead threshold for the adverse effects of lead on children had been identified and the Center for Disease Control (CDC) currently uses a reference value of 5 μg/dL. Chapter 8, by Akkus and Ozdenerol , provides a review of the recent literature on GIS-based studies that suggests that numerous environmental risk factors might be critical for lead exposure. New GIS-based studies are used in surveillance data management, risk analysis, lead exposure visualization, and community intervention strategies where geographically-targeted, specific intervention measures are taken.

Chapter 9, by Zahran and colleagues, evaluates atmospheric concentrations of soil and Pb aerosols, and blood lead levels (BLLs) in 367 839 children (ages 0–10) in Detroit, Michigan from 2001 to 2009 to test a hypothesized soil $\rightarrow$ air dust $\rightarrow$ child pathway of contemporary Pb risk. Atmospheric soil and Pb show near-identical seasonal properties that match seasonal variation in children's BLLs. Resuspended soil appears to be a significant underlying source of atmospheric Pb. A 1% increase in the amount of resuspended soil results in a 0.39% increase in the concentration of Pb in the atmosphere (95% CI, 0.28 to 0.50%). In turn, atmospheric Pb significantly explains age-dependent variation in child BLLs. Other things held equal, a change of 0.0069 μg/m3 in atmospheric Pb increases BLL of a child 1 year of age by 10%, while approximately 3 times the concentration of Pb in air (0.023 μg/m^3) is required to induce the same increase in BLL of a child 7 years of age. Similarly, a 0.0069 μg/m^3 change in air Pb increases the odds of a child <1 year of age having a BLL $\geq$ 5 μg/dL by a multiplicative factor of 1.32 (95% CI, 1.26 to 1.37). Overall, the resuspension of Pb contaminated soil explains observed seasonal variation in child BLLs.

Gestational exposure to several common agricultural pesticides can induce developmental neurotoxicity in humans, and has been associated with developmental delay and autism. In Chapter 10, Shelton and colleagues aimed to evaluate whether residential proximity to agricultural pesticides during pregnancy is associated with autism spectrum disorders (ASD) or developmental delay (DD) in the Childhood Autism Risks from Genetics and Environment (CHARGE) Study. The CHARGE study is a population-based case-control study of ASD, developmental delay (DD), and typical development. For 970 participants, commercial pesticide application data from the California Pesticide Use Report (1997-2008) were linked to the addresses during pregnancy. Pounds of active ingredient applied for organophophates, organochlorines, pyrethroids, and carbamates were aggregated within 1.25km, 1.5km, and 1.75km buffer distances from the home. Multinomial logistic regression was used to estimate the odds ratio (OR) of exposure comparing confirmed cases of ASD (n = 486) or DD (n = 168) with typically developing referents (n = 316). Approximately one-third of CHARGE Study mothers lived, during pregnancy, within 1.5 km (just under one mile) of an agricultural pesticide application. Proximity to organophosphates at some point during gestation was associated with a 60% increased risk for ASD, higher for 3rd trimester exposures [OR = 2.0, 95% confidence interval (CI) = (1.1, 3.6)], and 2ndtrimester chlorpyrifos applications: OR = 3.3 [95% CI = (1.5, 7.4)]. Children of mothers residing near pyrethroid insecticide applications just prior to conception or during 3rd trimester were at greater risk for both ASD and DD, with OR's ranging from 1.7 to 2.3. Risk for DD was increased in those near carbamate applications, but no specific vulnerable period was identified This study of ASD strengthens the evidence linking neurodevelopmental disorders with gestational pesticide exposures, and particularly, organophosphates and provides novel results of ASD and DD associations with, respectively, pyrethroids and carbamates.

In a longitudinal birth cohort study of inner-city mothers and children (Columbia Center for Children's Environmental Health), the authors of Chapter 11, Rauh and colleagues, have previously reported that prenatal exposure to chlorpyrifos (CPF) was associated with neurodevelopmental problems at 3 years of age. The goal of the current study was to estimate the relationship between prenatal CPF exposure and neurodevelopment among cohort children at 7 years of age. In a sample of 265 children, participants in a prospective study of air pollution, the authors measured prenatal CPF exposure using umbilical cord blood plasma (picograms/gram plasma) and 7-year neurodevelopment using the Wechsler Intelligence Scale for Children, 4th edition (WISC-IV). Linear regression models were used to estimate associations, with covariate selection based on two alternate approaches. On average, for each standard deviation increase in CPF exposure (4.61 pg/g), Full-Scale intelligence quotient (IQ) declined by 1.4% and Working Memory declined by 2.8%. Final covariates included maternal educational level, maternal IQ, and quality of the home environment. The authors found no significant interactions between CPF and any covariates, including the other chemical exposures measured during the prenatal period (environmental tobacco smoke and polycyclic aromatic hydrocarbons). The study reports evidence of deficits in Working Memory Index and Full-Scale IQ as a function of prenatal CPF exposure at 7 years of age. These findings are important in light of continued widespread use of CPF in agricultural settings and possible longer-term educational implications of early cognitive deficits.

Polybrominated diphenyl ethers (PBDEs) are widely used flame retardant compounds that are persistent and bioaccumulative and therefore have become ubiquitous environment contaminants. Animal studies suggest that prenatal PBDE exposure may result in adverse neurodevelopmental effects. Chapter 12, by Herbstman and colleagues, provides a longitudinal cohort initiated after 11 September 2001, including 329 mothers who delivered in one of three hospitals in lower Manhattan, New York. The authors examined prenatal PBDE exposure and neurodevelopment when their children were 12–48 and 72 months of age. They analyzed 210 cord blood specimens for selected PBDE congeners and assessed neurodevelopmental effects in the children at 12–48 and 72 months of age; 118, 117, 114, 104, and 96 children with available cord PBDE measurements were assessed at 12, 24, 36, 48, and 72 months, respectively. The authors used multivariate regression analyses to evaluate the associations between concentrations of individual PBDE congeners and neurodevelopmental indices. Median cord blood concentrations of PBDE congeners 47, 99, and 100 were 11.2, 3.2, and 1.4 ng/g lipid, respectively. After adjustment for potential confounders, children with higher concentrations of BDEs 47, 99, or 100 scored lower on tests of mental and physical development at 12–48 and 72 months. Associations were significant for 12-month Psychomotor Development Index (BDE-47), 24-month Mental Development Index (MDI) (BDE-47, 99, and 100), 36-month MDI (BDE-100), 48-month full-scale and verbal IQ (BDE-47, 99, and 100) and performance IQ (BDE-100), and 72-month performance IQ (BDE-100). This epidemiologic study demonstrates neurodevelopmental effects in relation to cord blood PBDE concentrations. Confirmation is needed in other longitudinal studies.

Animal and *in vitro* studies demonstrated a neurotoxic potential of brominated flame retardants, a group of chemicals used in many household and commercial products to prevent fire. Although the first reports of detrimental neurobehavioral effects in rodents appeared more than ten years ago, human data are sparse. As a part of a biomonitoring program for environmental health surveillance in Flanders, Belgium, Chapter 13, by Kiciński and colleagues, assessed the neurobehavioral function with the Neurobehavioral Evaluation System (NES-3), and collected blood samples in a group of high school students. Cross-sectional data on 515 adolescents (13.6-17 years of age) was available for the analysis. Multiple regression models accounting for potential confounders were used to investigate the associations between biomarkers of internal exposure to brominated flame retardants [serum levels of polybrominated diphenyl ether (PBDE) congeners 47, 99, 100, 153, 209, hexabromocyclododecane (HBCD), and tetrabromobisphenol A (TBBPA)] and cognitive performance. In addition, the authors investigated the association between brominated flame retardants and serum levels of FT3, FT4, and TSH. A two-fold increase of the sum of serum PBDE's was associated with a decrease of the number of taps with the preferred-hand in the Finger Tapping test by 5.31 (95% CI: 0.56 to 10.05, p=0.029). The effects of the individual PBDE congeners on the motor speed were consistent. Serum levels above the level of quantification were associated with an average decrease of FT3 level by 0.18 pg/mL (95% CI: 0.03 to 0.34, p=0.020) for PBDE-99 and by 0.15 pg/mL (95% CI: 0.004 to 0.29, p=0.045) for PBDE-100, compared with concentrations below the level of quantification. PBDE-47 level above the level of quantification was associated with an average increase of TSH levels by 10.1% (95% CI: 0.8% to 20.2%, p=0.033), compared with concentrations below the level of quantification. The authors did not observe effects of PBDE's on neurobehavioral domains other than the motor function. HBCD and TBBPA did not show consistent associations with performance in the neurobehavioral tests. This study is one of few studies and so far the largest one investigating the neurobehavioral effects of brominated flame retardants in humans. Consistently with experimental animal data, PBDE exposure was associated with changes in the motor function and the serum levels of the thyroid hormones.

Chapter 14, by Schmidt, provides an editorial look at recent research in the transgenerational effects of environmental exposures. The article describes how environmental toxins or stresses can affect offspring up to three generations later, and calls for more human studies to understand these effects more fully.

Chapter 15, by O'Reilly and Shimkin, concludes this book with another article taken from the WHO report on *Children's Health and the Environment*. This chapter makes the argument that in the fight to preserve children's environmental health, every level has a role to play, from member of the community to governments on all levels to researchers. The authors describe how to understand these different forces at work in the protection of children, and argue that a more widespread and concerted effort needs to be made to narrow gaps in the quality of care between nations so that every child has a chance at a healthy and productive life.

PART I

INTRODUCTION

CHAPTER 1

CHILDREN ARE NOT LITTLE ADULTS (EXCERPT FROM *CHILDREN'S HEALTH AND THE ENVIRONMENT: A GLOBAL PERSPECTIVE*)

P. J. LANDRIGAN AND A. GARG

1.1 INTRODUCTION

Children around the world today confront environmental hazards that were neither known nor suspected a few decades ago. More than 80,000 new synthetic chemical compounds have been developed over the past 50 years. Children are especially at risk of exposure to the 15,000 of these chemicals produced in quantities of 4,500 kg or more per year, and to the more than 2800 chemicals produced in quantities greater than 450,000 kg per year. These high-production volume (HPV) chemicals are those most widely dispersed in air, water, food crops, communities, waste sites and homes (1). Worldwide many thousands of deaths occur as a result of poi-

Reprinted with Permission from the WHO. Landrigan PJ and Garg A. "Children Are Not Little Adults." In Children's Health and the Environment: A Global Perspective, *edited by J. Pronczuk-Garbino, Geneva: World Health Organization, 2005.*

soning, with the vast majority being among children and adolescents after accidental exposure. Many hundreds of HPV chemicals have been tested for their potential human toxicity, but fewer than 20% have been examined for their potential to cause developmental toxicity to fetuses, infants, and children (1, 2).

Until about ten years ago, chemical exposure was principally a problem for children in the developed countries. However, it is becoming a problem in developing countries as hazardous industries relocate there as a consequence of globalization and in an effort to escape ever stricter labour and environmental laws in the developed countries.

In addition to the hazards of new chemicals, children worldwide confront traditional environmental hazards, including poor water quality and sanitation, ambient and indoor air pollution, vector-borne diseases, unintentional injuries, inadequate housing, and effects of climate variability and change.

1.2 CHILDREN'S UNIQUE SUSCEPTIBILITY

Children are highly vulnerable to environmental hazards for several reasons (3, 4)· Children have disproportionately heavy exposures to environmental toxicants. In relation to body weight, children drink more water, eat more food, and breathe more air than adults. Children in the first six months of life drink seven times as much water per kg of body weight and 1-5-year-old children eat 3-4 times more food per kg than the average adult. The air intake of a resting infant is proportionally twice that of an adult. As a result, children will have substantially heavier exposures than adults to any toxicants that are present in water, food, or air. Two additional characteristics of children further magnify their exposures: their hand-to-mouth behaviour, and the fact that they live and play close to the ground.

Children's metabolic pathways, especially in the first months after birth, are immature. Children's ability to metabolize, detoxify, and excrete many toxicants is different from that of adults. In some cases, children may actually be better able than adults to deal with some toxicants, e.g.

paracetamol. Commonly, however, they are less well able to deal with toxic chemicals and thus are more vulnerable to them.

Children undergo rapid growth and development, and their developmental processes are easily disrupted. The organ systems of infants and children change very rapidly before birth, as well as in the first months and years of life. These developing systems are very delicate and are not able to repair adequately damage caused by environmental toxicants. Thus, if cells in an infant's brain are destroyed by chemicals such as lead, mercury, or solvents, or if false signals are sent to the developing reproductive organs by endocrine disruptors, there is a high risk that the resulting dysfunction will be permanent and irreversible.

Because children generally have more future years of life than adults, they have more time to develop chronic diseases triggered by early exposures. Many diseases that are caused by toxicants in the environment require decades to develop. Many such diseases, including cancer and neurodegenerative diseases, are now thought to arise through a series of stages that require years or even decades from initiation to actual manifestation of disease. Carcinogenic and toxic exposures sustained early in life, including prenatal exposures, appear more likely to lead to disease than similar exposures encountered later.

1.3 DISEASES IN CHILDREN POSSIBLY LINKED TO ENVIRONMENTAL EXPOSURES

Children are exposed to a series of health risks from environmental hazards. Environment-related illnesses are responsible for more than 4.7 million deaths annually in children under the age of five (5). Both "basic" and traditional risks, such as unsafe water, poor sanitation, indoor air pollution, poor food hygiene, poor quality housing, inadequate waste disposal, vector-borne diseases and hazards that cause accidents and injuries, as well as "modern" environmental risks endanger children's health. Newly emerging environmental threats to the health of children derive from high levels of natural or man-made toxic substances in the air, water, soil and food chain, global climate change and ozone depletion, electromagnetic radiation and contamination by persistent organic pollutants and chemicals that disrupt endocrine functions.

1.4 ENVIRONMENTAL HEALTH HAZARDS

ʻThe risks to children in their everyday environments are numerous. But there are six groups of environmental health hazards that cause the bulk of environmentally related deaths and disease among children (12), as outlined below.

1.4.1 HOUSEHOLD WATER SECURITY

Contaminated water is the cause of many life-threatening diseases including diarrhoea, the second biggest child-killer in the world. Diarrhoea is estimated to cause 1.3 million child deaths per year-about 12% of total deaths of children under five in developing countries. Around the world, both biological disease agents and chemical pollutants are compromising the quality of drinking-water. Water contamination can spread diseases such as hepatitis B, dysentery, cholera and typhoid fever. High levels of arsenic, lead or fluoride may lead to both acute and chronic diseases in children.

1.4.2 SUBCLINICAL TOXICITY

A critically important intellectual step in the development of understanding of children's special susceptibility to chemical toxins has been the recognition that environmental toxins can cause a range of adverse effects in children. Some of these effects are clinically evident, but others can be discerned only through special testing and are not evident on the standard examination-hence the term "subclinical toxicity." The underlying concept is that there is a dose-dependent continuum of toxic effects, in which clinically obvious effects have their subclinical counterparts (6, 1).

The concept of subclinical toxicity has its origins in the pioneering studies of lead toxicity in clinically asymptomatic children undertaken by Herbert Needleman and colleagues (8). Needleman et al. showed that children's exposure to lead could cause decreased intelligence and altered behaviour, even in the absence of clinical symptoms of lead toxicity. The subclinical toxicity of lead in children has subsequently been confirmed in prospective epidemiological studies (9). Similar subclinical neurotoxic effects have been documented in children exposed in the womb to polychlorinated biphenyls (PCBs) (10) and to methylmercury (11).

1.4.3 HYGIENE AND SANITATION

Globally, 2.4 billion people, most of them living in periurban or rural areas in developing countries, do not have access to any type of sanitation facilities. Coverage estimates for 1990-2000 show that little progress was made during this period in improving coverage. The lowest levels of service coverage are found in Africa and Asia, where 48% and 31% of the rural populations, respectively, do not have these services. Examples of sanitation-related diseases include cholera, typhoid, schistosomiasis, and trachoma-a disease that causes irreversible blindness, and currently affects about 6 million people, with another 500 million at risk of the disease.

1.4.4 AIR POLLUTION

Air pollution is a major environment-related health threat to children and a risk factor for both acute and chronic respiratory disease as well as a range of other diseases. Around two million children under five die every year from acute respiratory infections (ARI) aggravated by environmental hazards. Indoor air pollution is a major causal factor for ARI deaths in rural and urban areas of developing countries. Outdoor air pollution, mainly from traffic and industrial processes, remains a serious problem in cities throughout the world, particularly in the ever-expanding megacities of developing countries. A major problem is the continuing use of lead in petrol

(13). It is estimated that a quarter of the world's population is exposed to unhealthy concentrations of air pollutants such as particulate matter, sulfur dioxide, and other chemicals.

1.4.5 DISEASE VECTORS

Numerous vector-borne diseases affect children's health. Their impact varies in severity. Malaria is particularly widespread and dangerous, existing in 100 countries and accounting for more than 800,000 deaths annually, mostly in children under five. Schistosomiasis is a water-borne disease that mainly affects children and adolescents, and is endemic in 74 developing countries. Japanese encephalitis occurs only in south and south-east Asia, where it is linked with irrigated rice production ecosystems. The annual number of clinical cases is estimated at about 40,000. Some 90% of cases are in children in rural areas, and 1 in 5 of these children dies. Annual mortality due to dengue is estimated at around 13,000; more than 80% of these deaths occur in children.

1.4.6 CHEMICAL HAZARDS

As a result of the increased production and use of chemicals, myriad chemical hazards are nowadays present in children's homes, schools, playgrounds and communities. Chemical pollutants are released into the environment by uncontrolled industries or through leakage from toxic waste sites. In 2002 about 350,000 people died as a result of poisoning, and 46,000 were children and adolescents exposed accidentally (www.who.int/evidence/bod). Pesticides, cleaners, kerosene, solvents, pharmaceuticals and other products unsafely stored or used at home are the most common causes of acute toxic exposure. Some result in life-threatening poisoning. Chronic exposure to a number of persistent environmental pollutants is linked to damage to the nervous and immune systems and to effects on reproductive function and development, as exposure occurs during periods of special susceptibility in the growing child or adolescent.

Children are quite vulnerable to the neurotoxic effects of lead in paint and air, which may reduce their intelligence and cause learning disabilities and behavioural problems, in particular reduced attentiveness. They are also vulnerable to the developmental effects of mercury released into the environment or present as a food contaminant. Another problem is asbestos, which is used extensively as an insulator in the construction industry (14).

1.4.7 INJURIES AND ACCIDENTS

In 2000 an estimated 685,000 children under the age of 15 were killed by an unintentional injury, accounting for approximately 20% of all such deaths worldwide. Unintentional injuries are among the ten leading causes of death for this age group. Worldwide, the leading causes of death from unintentional injury among children are road traffic injuries and drowning, accounting for 21% and 19%, respectively. Unintentional injuries among children are a global problem, but children and adolescents in certain regions of the world are disproportionately affected. It is estimated that 98% of all unintentional injuries in children occur in low- and middle-income countries, and 80% of all childhood deaths from unintentional injuries occur in the African, South-East Asian and Western Pacific regions.

1.4.8 CHRONIC EFFECTS OF ENVIRONMENTAL HAZARDS

Exposure to environmental hazards is known or suspected to be responsible for a series of acute and chronic diseases that, in the industrialized countries, have replaced infectious diseases as the principal causes of illness and death in childhood.

Urbanization and pervasive poverty in developing countries aggravate both "basic" and "modern" health risks. Developing countries therefore face a double burden in paediatric environmental health.

The chronic diseases represent a "new paediatric morbidity," and include the following.

1.4.8.1 ASTHMA

Asthma prevalence among children under 18 years of age has more than doubled over the past decade in many industrialized and developed countries. This increase is particularly evident in urban centres, where asthma has become the leading cause of children's admissions to hospital and of school absenteeism (15, 16).

Ambient air pollutants, especially ground-level ozone and fine particulates from automobile exhausts, appear to be important triggers of asthma. Asthma incidence declines when levels of these pollutants drop (17). Indoor air pollution, including use of open fires for cooking and heating, insect dust, mites, moulds and environmental tobacco smoke are additional triggers.

Sharp discrepancies in asthma by socioeconomic and racial or ethnic status have been noted in certain countries. In New York City, hospital admission rates for asthma are 21 times higher in the poorest communities than in the wealthiest ones (18). Globally, the International Study of Asthma and Allergies in Childhood (ISAAC) demonstrated a wide range in rates for symptoms of asthma. Up to 15-fold differences were found between countries, with a range of 2.1% to 4-4% in Albania, China, Greece, Indonesia, Romania and the Russian Federation, and 29.1% to 32.2% in Australia, New Zealand, Ireland and the United Kingdom (19).

1.4.8.2 CHILDHOOD CANCER

The reported incidence of cancer among children under 18 years of age in the United States has increased substantially in the past 20 years (20). Indeed, childhood cancer incidences around the world are on the rise. Industrialized countries have succeeded in bringing the death rates from childhood cancer down, thanks to improved treatment. Still, in the United States, the incidence of acute lymphoblastic leukaemia (ALL), the most common childhood cancer, increased by 27.4% from 1973 to 1990, from 2.8 cases per 100000 children to 3.5 per 100000. Since 1990, ALL incidence has declined in boys in the United States, but continues to rise in girls. Between 1973 and 1994, incidence of primary brain cancer (glioma) increased by 39.6%, with nearly equal increases in boys and girls (21). In young white men, 20-39 years of age, although not in black men, incidence of testicular cancer increased by 68%. The causes of these increases are not known. In tropical Africa, Burkitt lymphoma is the most common childhood malignancy. However, in Nigeria there was a marked decrease in the relative frequency of Burkitt lymphoma from 37.1% in the period 1973 to 1990 to 19-4% between 1991 and 1999 which was seen partly as a consequence of improved living conditions and greater control of malaria (22). The risk of liver cancer is influenced by a number offactors; persistent infection with the hepatitis B or C virus is strongly associated with this kind of malignancy. Aflatoxins, one of the most potent mutagenic and carcinogenic mycotoxins, also represent a major risk factor for hepatocellular carcinoma, especially in highincidence areas, i.e. south-east Asia and parts of Africa. Immunization against hepatitis B or protection against hepatitis C and reduced aflatoxin exposure would reduce the risk for liver cancer in these populations (23, 24)·

1.4.8.3 LEAD POISONING

Countries that have succeeded in removing lead from petrol have accomplished a major public health goal that greatly benefits all citizens, especially children. There have been significant studies of lead in blood before and after elimination of lead from petrol in a number of industrialized countries. These studies show conclusively a direct relationship between lead in petrol and lead levels in children's blood. Still, even with the tremendous improvement in blood lead levels in countries that have removed lead from petrol, there are other sources of lead exposure. For example, in the USA, despite a 94% decline in blood lead levels since 1976, an estimated 930,000 preschool children in the United States still have elevated blood lead levels (10µg/dl or above) and suffer from lead toxicity (25). These children are at risk of diminished intelligence, behavioural disorders, failure at school, delinquency, and diminished achievement (9). Rates of lead poisoning are highest in poor children from disadvantaged groups in urban centres. New immigrants to the United States are often at high risk, because they tend to live in poor-quality housing, are not aware of the dangers of lead-based paint, and may bring medications or cosmetics containing lead from their home countries (25).

In the industrially developed countries, consumption of lead has decreased sharply in the past two decades. This reduction reflects the phasing out of leaded petrol and decreases in industrial use of lead (26). Major reductions in human exposure and in population blood lead levels have resulted. By contrast, in countries undergoing transition to industrialization, lead use in petrol as well as in industry remains widespread, environmental contamination may be intense, and blood lead levels in workers as well as in residents of communities near polluting industries have been reported to be dangerously elevated (27-30). A study conducted in Romania examined lead levels in children from a polluted municipality: only six children out of 42 had blood lead levels below 10μg/dl (27). A devastating experience in Trinidad and Tobago demonstrated the impact of lead exposure on child health, in this case from battery recycling. A six-year-old boy died from acute lead poisoning, with a blood lead level above 140μg/dl, and many children had to undergo chelating therapy and suffered permanent damage from exposure to extremely high lead levels.

1.4.8.4 DEVELOPMENTAL DISORDERS

Developmental disorders, including autism, attention deficit disorder, dyslexia and mental retardation affect 5-8% of the 4 million children born each year in the United States (31). The causes are largely unknown, but exposure to lead, mercury, PCBs, certain pesticides and other environmental neurotoxicants are thought to contribute. An expert committee convened by the US National Academy of Sciences concluded in July 2000 that 3% of all developmental disorders in American children are the direct consequence of toxic environmental exposures, and that another 25% are the result of interactions between environmental factors and individual children's susceptibility (32).

1.4.8.5 ENDOCRINE DISRUPTION

Endocrine disruptors are chemicals in the environment that have the capacity to interfere with the body's hormonal signalling system. The effects of these chemicals have been well documented in experimental animals exposed in the laboratory as well as in wildlife populations in contaminated ecosystems such as the Great Lakes in North America.

While data on the human health effects of endocrine disruptors are still scant, it would appear that the embryo, fetus and young child are at greatest risk of adverse consequences following exposure to these chemicals, because the human reproductive and endocrine systems undergo complex development in fetal life and thus are highly vulnerable to toxic influences. It is hypothesized, but not proven, that endocrine·disrupting compounds may be responsible, at least in part, for an increased incidence of testicular cancer, a reported doubling in incidence of hypospadias (33) and the increasingly early onset of puberty in young girls.

1.5 THE INTERNATIONAL RESPONSE TO CHILDREN'S ENVIRONMENTAL HEALTH

The first major international development in children's environmental health was the Declaration of the Environment Leaders of the Eight on Children's Environmental Health, issued in Miami in 1997 by the group of highly industrialized nations, the so-called G-8 (34). The Miami Declaration expressed the commitment of these nations to children's environmental health and included specific commitments to remove lead from petrol, to improve air quality, and to improve the quality of drinking-water. The Declaration also called for improvements in the scientific risk assessments that underpin environmental regulations to explicitly incorporate children, and set forth international cooperation to do further research on endocrine-disrupting chemicals. This Declaration has catalysed developments in many international organizations and nongovernmental organizations (NGOs).

In 2002, the World Health Organization (WHO) launched an initiative to improve environmental protection of children, reflecting its major thrust at the World Summit on Sustainable Development, which took place in Johannesburg, South Africa. "Our top priority in health and development must be investing in the future-in children and the young—a group that is particularly vulnerable to environmental hazards," stated WHO's then Director-General, Dr Gro Harlem Brundtland. She set forth the Healthy Environments for Children Alliance which many international organizations, nations, NGOs and businesses have responded to and have begun to put into action.

A WHO working group on children's environmental health has been active in bringing together participants from developed as well as developing countries since early 2000. The United Nations Environment Programme (UNEP), the World Bank, and the United Nations Children's Fund (UNICEF) have joined in partnership with WHO in these efforts.

The international community of NGOs is becoming active in children's environmental health. An umbrella organization, the International Network of Children's Environmental Health & Safety (INCHES), was formed to link grassroots organizations in various countries. INCHES, in conjunction with a US NGO, the Children's Environmental Health Network (CEHN) hosted a global conference on children's environmental health in Washington in September 2001 to raise international interest in children's environmental health. This was followed by a WHO conference in Bangkok, Thailand, which considered environmental threats to the health of children in South-East Asia and the Western Pacific (see resulting statement below) (35). The Pan American Health Organization has developed a strategy to improve environmental health of children in the Americas and launched a regional workshop on children's environmental health in 2003.

THE BANGKOK STATEMENT

A pledge to promote the protection of Children's Environmental Health
We, the undersigned scientists, doctors and public health professionals, educators, environmental health engineers, community workers and repre-

sentatives from a number of international organizations, from governmental and non-governmental organizations in South East Asian and Western Pacific countries, have come together with colleagues from different parts of the world from 3 to 7 March 2002 in Bangkok, Thailand, to commit ourselves to work jOintly towards the promotion and protection of children's health against environmental threats.

Worldwide, it is estimated that more than one quarter of the global burden of disease (GBD) can be attributed to environmental risk factors. Over 40% of the environmental disease burden falls on children under 5 years of age, yet these constitute only 10% (of the world population. The environmental burden of pediatric disease in Asia and the Pacific countries is not well recognized and needs to be Quantified and addressed.

We recognize

That a growing number of diseases in children have been linked to environmental exposures. These range from the traditional waterborne, foodborne and vector-borne diseases and acute respiratory infections to asthma, cancer, injuries, arsenicosis, fluorosis, certain birth defects and developmental disabilities.

That environmental exposures are increasing in many countries in the region; that new emerging risks are being identified; and that more and more children are being exposed to unsafe environments where they are conceived and born, where they live, learn, play, work and grow. Unique and permanent adverse health effects can occur when the embryo, fetus, newborn, child and adolescent (collectively referred to as "children" from here onwards) are exposed to environmental threats during early periods of special vulnerability.

That in developing countries the main environmental health problems affecting children are exacerbated by poverty, illiteracy and malnutrition, and include: indoor and outdoor air pollution, lack of access to safe water and sanitation, exposure to hazardous chemicals, accidents and injuries. Furthermore, as countries industrialize, children become exposed to toxicants commonly associated with the developed world, creating an additional environmental burden of disease. This deserves special attention from the industrialized and developing countries alike.

That environmental hazards arise both from anthropogenic and natural sources (e.g. plant toxins, fluoride, arsenic, radiations), which separately and in combination can cause serious harm to children.

That restoring and protecting the integrity of the life-sustaining systems of the earth are integral to ensuring children's environmental health now and in the future. Therefore, addressing global changes such as human population growth, land and energy use patterns, habitat destruction, biodiversity loss and climate change must be part of efforts to promote children's environmental health.

That despite the rising concern of the scientific community and the education and social sectors about environmental threats to children's health and development, progress has been slow and serious challenges still remain.

That the health, environment and education sectors must take concerted action at all levels (local, national, global), together with other sectors, in serious efforts to enable our countries to assess the nature and magnitude of the problem, identify the main environmental risks to children's health and establish culturally appropriate monitoring, mitigation and prevention strategies.

We affirm

That the principle "children are not little adults" requires full recognition and a preventive approach. Children are uniquely vulnerable to the effects of many chemical, biological and physical agents. All children should be protected from injury, poisoning and hazards in the different environments where they are born, live, learn, play, develop and grow to become the adults of tomorrow and citizens in their own right.

That all children should have the right to safe, clean and supportive environments that ensure their survival, growth, development, healthy life and well-being. The recognition of this right is especially important as the world moves towards the adoption of sustainable development practices.

That it is the responsibility of community workers, local and national authorities and policymakers, national and international organizations, and all professionals dealing with health, environment and education issues to ensure that actions are initiated, developed and sustained in all countries to promote the recognition, assessment and mitigation of physical, chemical and biological hazards, and also of social hazards that threaten children's health and quality of life.

We commit ourselves

To developing active and innovative national and international networks with colleagues, in partnership with governmental, nongovernmental and international organizations for the promotion and protection of children's environmental health, and urge WHO to support our efforts in all areas, especially in the following four:

1. Protection and Prevention—To strengthen existing programmes and initiate new mechanisms to provide all children with access to clean water and air, adequate sanitation, safe food and appropriate shelter:

- Reduce or eliminate environmental causes and triggers of respiratory diseases and asthma, including exposure to indoor air pollution from the use of biomass fuels and environmental tobacco smoke.
- Reduce or eliminate exposure to toxic metals such as lead, mercury and arsenic, to fluoride, and to anthropogenic hazards such as toxic wastes, pesticides and persistent organic pollutants.
- Reduce or eliminate exposure to known and suspected anthropogenic carcinogens, neurotoxicants, developmental and reproductive toxicants, immunotoxicants and naturally occurring toxins.
- Reduce the incidence of diarrhoeal disease through increased access to safe water and sanitation and promotion of initiatives to improve food safety.
- Reduce the incidence of accidents, injuries and poisonings, as well as exposure to noise, radiation, microbiological and other factors by improving all environments where children spend time, in particular at home and at school.
- Commit to international efforts to avert or slow global environmental changes, and also take action to lessen the vulnerability of populations to the impact of such changes.

2. Health Care and Research—To promote the recognition, assessment and study of environmental factors that have an impact on the health and development of children:

- Establish centres to address issues related to children's environmental health.
- Develop and implement cooperative multidisciplinary research studies in association with centres of excellence, and promote the collection of harmonized data and their dissemination.

- Incorporate children's environmental health into the training for health care providers and other professionals, and promote the use of the environmental history.
- Seek financial and institutional support for research, data collection, education, intervention and prevention programmes.
- Develop risk assessment methods that take account of children as a special risk group.

3. Empowerment and Education—To promote the education of children and parents about the importance of their physical environment and their participation in decisions that affect their lives, and to inform parents, teachers and caregivers and the community in general on the need and means to provide a safe, healthy and supportive environment to all children:

- Provide environmental health education through healthy schools and adult education initiatives.
- Incorporate lessons on health and the environment into all school curricula.
- Empower children to identify potential risks and solutions.
- Impart environmental health expertise to educators, curriculum designers and school administrators.
- Create and disseminate to families and communities culturally relevant information about the special vulnerability of children to environmental threats and practical steps to protect children.
- Teach families and the community to identify environmental threats to their children, to adopt practices that will reduce risks of exposure and to work with local authorities and the private sector in developing prevention and intervention programmes.

4. Advocacy—To advocate and take action on the protection and. promotion of children's environmental health at all levels,including political,administralive and community levels:

- Use lessons learned to prevent environmental illness in children, for example by promoting legislation for the removal of lead from all petrol, paints, water pipes and ceramics, and for the provision of smoke-free environments in all public buildings.
- Sensitize decision makers to the results of research studies and observations of community workers and primary health care providers that need to be accorded high priority to safeguard children's health.
- Promote environmental health policies that protect children.

- Raise the awareness of decision-makers and potential donors about known environmental threats to children's health and work with them and other stakeholders to allocate necessary resources to implement interventions.
- Work with the media to disseminate information on core children's environmental health issues and locally relevant enironmental health problems and potential solutions.

For all those concerned about the environmental health of children, the time to translate knowledge into action is now.

Bangkok, 7 March 2002

REFERENCES

1. US Environmental Protection Agency. Chemicals-in-commerce information system. Chemical update system database, 1998. Washington, DC, July 1997.
2. National Academy of Sciences. Toxicity testing needs and priorities. Washington, DC, National Academy Press, 1984.
3. US Environmental Protection Agency. Office of Pollution Prevention and Toxic Substances. Chemical hazard data availability study: What do we really know about the safety of high production volume chemicals? US EPA, Washington, DC, 1998.
4. National Academy of Sciences. Pesticides in the diets of infants and children. Washington, DC, National Academy Press, 1993.
5. World Health Organization. Healthy environments for children. World Health Organization, Geneva, 2002 (WHO/SDE/PHE/02.05).
6. Landrigan PJ, Suk WA, Amler RW. Chemical wastes, children's health and the Superfund Basic Research Program. Environmental Health Perspectives, 1999, 107:423-427.
7. Landrigan PJ. The toxicity of lead at low dose. British Journal of Industrial Medicine, 1989, 46:593-596.
8. Needleman H L, Gunnoe C, Leviton A. Deficits in psychological and classroom performance of children with elevated dentine lead levels. New England Journal of Medicine, 1979, 300:689-695.
9. Bellinger D et al. Longitudinal analysis of prenatal and postnatal lead exposure and early cognitive development. New England Journal of Medicine, 1987, 316:1037-1043.
10. Jacobson JL, Jacobson SW, Humphrey HEB. Effects of in utero exposure to polychlorinated biphenyls and related contaminants on cognitive functioning in young children. Journal of Pediatrics, 1990, 116:38-45.
11. Grandjean Pet al. Cognitive deficit in 7-year-old children with prenatal exposure to methylmercury. Neurotoxicology and Teratology, 1997, 19:417-428.

12. World Health Organization. Healthy environments for children. Initiating an alliance for action. World Health Organization, Geneva, 2002 (WHO/SDE/PHE/02.06).

13. Landrigan PJ, Boffetta P, Apostoli P. The reproductive toxicity and carcinogenicity of lead: a critical review. American Journal of Industrial Medicine, 2000, 38:231-243.

14. Nicholson WJ, Perkel G, Sclikoff IJ. Occupational exposure to asbestos: population at risk and projected mortalitY-1980-203°. American Journal of Industrial Medicine, 1982, 3:259-311.

15. Mannino DM et al. Surveillance for asthma-United States, 1960-1996. Morbidity and Mortality Weekly Report, 1998,47(55-1):1-28.

16. Centers for Disease Control and Prevention. Asthma mortality and hospitalization among children and young adults-United States, 1980-1993. Morbidity and Mortality Weekly Report, 1996, 45:350-353.

17. Friedman MS et al. Impact of changes in transportation and commuting behaviors during the 1996 summer Olympic Games in Atlanta on air quality and childhood asthma. Journal of the American Medical Association, 2001, 285:897-905.

18. Claudio Let al. Socioeconomic factors and asthma hospitalization rates in New York City. Journal of Asthma, 1999, 36:343-350.

19. Children's health and environment: a review of evidence. Copenhagen, WHO Regional Office for Europe, 2002 (Environmental Issue Report, No. 29).

20. Gurney JG et al. Trends in cancer incidence among children in the U.S. Cancer, 1996,76:532-4l.

21. Legler J M et al. Brain and other central nervous system cancers: recent trends in incidence and mortality. Journal of the National Cancer Institute, 1999, 91:1382-139°.

22. Ojesina AI, Akang EEU, Ojemakinde KO. Decline in the frequency of Burkitt's lymphoma relative to other childhood malignancies in Ibadan, Nigeria. Annals of Tropical Paediatrics, 2002, 22:159-163.

23. Henry SH, Bosch FX, Bowers Jc. Aflatoxin, hepatitis and worldwide liver cancer risks. Advances in Experimental Medicine and Biology, 2002, 504:229-233.

24. Montesano R, Hainaut P, Wild CPo Hepatocellular carcinoma: from gene to public health. Journal of the National Cancer Institute, 1997, 89:1844-1851.

25. Centers for Disease Control and Prevention. Lead poisoning-update. Morbidity and Mortality Weekly Report, 1997,46:141-146.

26. Hernberg S. Lead poisoning in a historical perspective. AmericanJournal of Industrial Medicine, 2000, 38:244-254.

27. Bindea V. Blood lead levels in children from Baia Mare, Romania. In: Proceedings of the International Conference on Lead Exposure, Reproductive Toxicity and Carcinogenicity, Gargagno, Italy, 7-9 June 1999, Lyon, International Agency for Research on Cancer, 1999.

28. Bulat P et al. Occupational lead intoxication in lead smelter workers. In: Proceedings of the International Conference on Lead Exposure, Reproductive Toxicity and Carcinogenicity, Gargagno, Italy, 7-9 June 1999, Lyon, International Agency for Research on Cancer, 1999.

29. Chatterjee A et al. Pollution from a lead smelter in a residential area of Calcutta. In: Proceedings of the International Conference on Lead Exposure, Reproductive

Toxicity and Carcinogenicity, Gargagno, Italy, 7-9 June 1999, Lyon, International Agency for Research on Cancer, 1999.

30. Koplan J. Hazards of cottage and small industries in developing countries. American Journal of Industrial Medicine, 1996, 30:123-124.

31. Weiss B, Landrigan P. The developing brain and the environment: an introduction. Environmental Health Perspectives, 2000, 108 (SupPI.):373-374·

32. National Academy of Sciences. Scientific frontiers in developmental toxicology and risk assessment. Washington, DC, National Academy Press, 2000.

33. Paulozzi LJ, Erickson JD, Jackson RJ. Hypospadias trends in two US surveillance systems. Pediatrics, 1997, 100:831-834.

34. Declaration of the Environment Leaders of the Eight on Children's Environmental Health. (http://www.library.utoronto.ca/g7 /environment/1997miami/children.html).

35. nternational Conference on Environmental Threats to the Health of Children: Hazards and Vulnerability. Bangkok, 3-7 March 2002 (http://www.who.int/docstore/peh/ceh/Bangkok/bangkokconf.htm).

CHAPTER 2

WINDOWS OF SUSCEPTIBILITY TO ENVIRONMENTAL EXPOSURES IN CHILDREN (EXCERPT FROM *CHILDREN'S HEALTH AND THE ENVIRONMENT: A GLOBAL PERSPECTIVE*)

S. G. SELEVAN, C. A. KIMMEL, AND P. MENDOLA

Children have a unique susceptibility to chemical, biological and physical environmental threats. Their tissues and organs grow rapidly, developing and differentiating until maturity. The developmental and growth processes in the fetus, infant, child, and adolescent can define periods of varying vulnerability to environmental toxicants. Furthermore, the exposure patterns and behaviours of children are very different from those of adults, and may result in greater exposures.

A large number of anatomical, biochemical, and physiological changes occur from early intrauterine life through adolescence. These maturational processes may be altered by physical, biological, and chemical environmental factors at various points in time. Furthermore, the changes with maturation may themselves substantially affect the absorption, distribu-

Reprinted with Permission from the WHO. Selevan G, Kimmel CA, and Mendola P. "Windows of Susceptibility to Environmental Exposures in Children." In Children's Health and the Environment: A Global Perspective, *edited by J. Pronczuk-Garbino, Geneva: World Health Organization, 2005.*

tion, metabolism, and elimination of chemicals present in the environment. The younger and less mature the child, the more different its response may be from that of an adult. Additionally, recent evidence suggests that exposure of the embryo, fetus or young child may affect the onset of diseases in adulthood, e.g. cardiovascular and neurodegenerative diseases or cancer. These susceptible developmental periods, called "windows of susceptibility," are times when a number of systems, including the endocrine, reproductive, immune, respiratory, and nervous systems, may be particularly sensitive to certain chemicals and physical factors. Therefore, a new "child-centred" approach to research, risk assessment and risk management is necessary to identify, understand, control, and prevent childhood or adult diseases of environmental origin.

2.1 TIMING OF EXPOSURE

Not only the level of exposure, but also its timing, may ultimately affect the health outcomes observed in children. During the highly susceptible periods of organ formation, timing of exposure is extremely important. For example, in the case of prenatal exposures, the same type of exposure occurring at different times is likely to produce a varied spectrum of malformations with the specific outcomes observed dependent on the organ system(s) most vulnerable at the time of exposure (Figure 1) (1).

In addition to highly sensitive windows for morphological abnormalities (birth defects), there are also time windows important for the development of physiological defects and morphological changes at the tissue, cellular and subcellular levels (Figure 2) (2). An adaptation of the scheme in Figure 2 shows the broader range of potential adverse outcomes from exposures during preconceptional, prenatal, and postnatal development (Figure 3). Many of the existing data are related to preconceptional and prenatal exposures (3). Data on prenatal exposures are based mainly on studies of maternal exposure to pharmaceuticals (e.g. diethylstilbestrol, thalidomide) and parental alcohol use, smoking, and occupational exposures. Information on critical windows for exposure during the postnatal period is scarce. Postnatal exposures have been examined in detail for only a few environmental agents, including lead, mercury, some pesticides, and radiation.

FIGURE 1: Syndromes of malformation

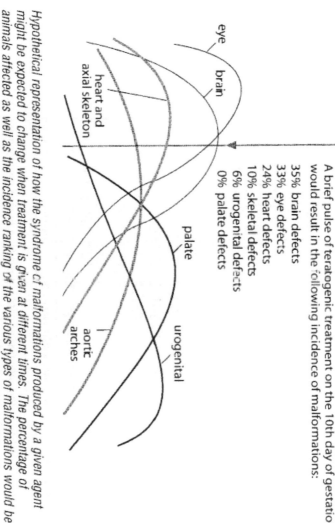

A brief pulse of teratogenic treatment on the 10th day of gestation would result in the following incidence of malformations:

35% brain defects
33% eye defects
24% heart defects
10% skeletal defects
6% urogenital defects
0% palate defects

eye

brain

heart and
axial skeleton

palate

aortic
arches

urogenital

Hypothetical representation of how the syndrome of malformations produced by a given agent might be expected to change when treatment is given at different times. The percentage of animals affected as well as the incidence ranking of the various types of malformations would be different if treatment were given on day 12 or 14, for example. Reprinted from (1) with permission of University of Chicago Press.

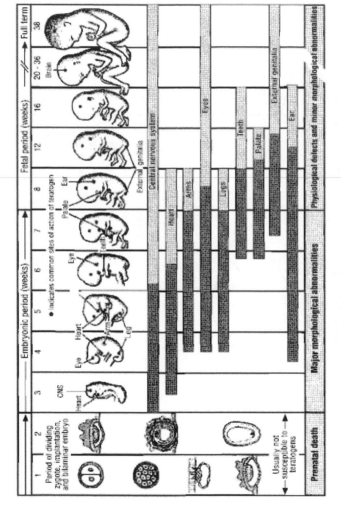

Schematic illustration of the sensitive or critical periods in human development. Dark grey denotes highly sensitive periods; light grey indicates stages that are less sensitive to teratogens. Reprinted from (2) with permission of W.B. Saunders Co.

FIGURE 2: Critical periods in human development

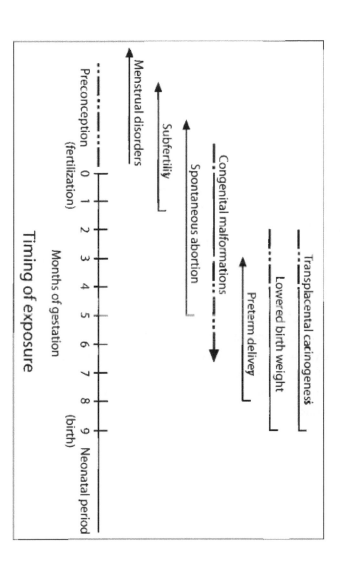

FIGURE 3: Reproductive outcomes and timing of maternal exposure

Solid lines indicate the most probable timing of exposure for a particular outcome; dotted lines indicate less probable but still possible timing of exposure. Arrows suggest that a defined cut-off point for exposure to a specific outcome is not known. Reprinted from (3) with permission of Lippincott, Williams and Wilkins.

Developmental exposures may result in health effects observed:

- prenatally and at birth, such as spontaneous abortion, stillbirth, low birth weight, small size for gestational age, infant mortality, and malformation (3-6);
- in childhood, such as asthma, cancer, neurological and behavioural effects (7-10);
- at puberty, such as alterations in normal development and impaired reproductive capacity (11, 12);
- in adults, such as cancer, heart disease, and degenerative neurological and behavioural disorders (13-16).

In 1999, a multidisciplinary group reviewed the data available on preconceptional, prenatal, and postnatal developmental exposures and the subsequent outcomes, looking in detail at the respiratory, immune, reproductive, nervous, cardiovascular and endocrine systems, as well as general growth and cancer. Clear limitations were found in the data available on developmental toxicity (5).

Animal studies done for regulatory testing purposes are well controlled and may include extended periods to simulate long-term human exposure. In humans, the patterns of exposure are much more variable and must be carefully evaluated with systematic studies of environmental exposures and developmental effects. Early evidence of the effects of environmental exposures on parents and children came from data on individuals with atypical exposures (e.g. industrial or environmental accidents, poisoning). These studies of atypical exposures provided information about associations between exposures and outcomes but were not usually sufficient for extrapolation because of limitations in the measurement of exposure and uncertainty as to whether exposure occurred at a biologically plausible time. More recent studies have attempted to better document exposures using environmental and biological sampling as well as measuring the level and timing of exposure.

Exposure issues vary according to the time at which they occur: before conception, prenatally, or postnatally. The preconceptional exposures of concern may occur acutely prior to conception or result from an increased body burden in either parent accumulated over a long period of exposure. Prenatally, exposures often change throughout pregnancy, e.g. if a woman reduces her alcohol consumption, quits smoking, or avoids using medicines. In addition to these variations, the altered absorption, distribution, metabolism, and excretion of chemicals during pregnancy result in changes in internal dosing (Table 1) (17). For example, pregnant women have an increased cardiac output and pulmonary function (they exchange about 72% more air over 8 hours at rest: 5000 litres in pregnant women versus 2900 litres in nonpregnant women) (18). In practice, studies of prenatal exposures in humans are difficult, as they are typically based upon one or a few measurements of exposure, which are used to estimate exposure for the entire pregnancy.

TABLE 1: Physiological and toxicokinetic changes during pregnancy[a]

Parameter	Physiological Change	Toxicokinetic Change
Absorption		
Gastric emptying time	Increased	Absorption increased
Intestinal motility	Decreased	Absorption increased
Pulmonary function	Increased	Pulmonary exposure increased
Cardiac output	Increased	Absorption increased
Blood flow to skin	Increased	Absorption increased
Dermal hydration	Increased	Absorption +/-
Metabolism		
Hepatic metabolism	+/-	Metabolism +/-
Extrahepatic metabolism	+/-	Metabolism +/-
Plasma proteins	Decreased	Metabolism +/-
Excretion		
Renal blood flow	Increased	Increased renal elimination
Glomerular filtration rate	Increased	Increased renal elimination
Pulmonary function	Increased	Increased pulmonary elimination
Plasma proteins	Decreased	Elimination +/-

[a] *Modified from Silvaggio & Mattison (17).*

Difficulties also arise in studies of exposure in children, as these can vary enormously over time. At different developmental stages, the biology, behaviour, settings, and activities of children can result in variable exposures which may also be quite different from those of adults in the same environment (Table 2). For example, the ratio of surface area to body mass in infants is approximately 2.7 times that in adults; the respiratory minute ventilation rate is more than 65 times greater (17), and consumption of drinking-water more than twice that of adults on a body weight basis (19). Even over the relatively short time span of childhood, exposure levels can vary widely.

2.1.1 WINDOWS OF SUSCEPTIBILITY

Because of the variety of biological factors mentioned above, there is a greater potential for adverse health effects in children than in adults. Children are still developing in many ways and may be more vulnerable. They may be less able to avoid exposure because of immature detoxification mechanisms. Differences in metabolism, size, and behaviour may mean that they have higher levels of exposure than adults in the same environment.

It is, therefore, crucial to identify and understand the importance of the "windows of susceptibility" in children and the relationships between exposures and developmental outcomes. This requires knowing the key periods of susceptibility for specific outcomes. Information on these critical windows has been compiled in a number of reports (20). However, one of the main constraints in identifying critical windows is the limited information on the exact timing and sensitivity of various developmental stages. This underscores the importance of collecting case-specific information and detailed exposure assessments in children. For cancer, the situation is unique: tumours may be induced in a wide variety of systems at many highly vulnerable developmental stages.

Several issues that are closely related to the windows of susceptibility in children require further attention.

- The consequences of developmental exposures that are manifested in adulthood and old age should be considered: the potential cascade of events that might result in health effects in the adult is poorly understood. For example, if intrauterine growth restriction (I UGR) is associated with exposure to a particular agent, does it have the same later-life effects as nutritionally induced IUGR?

TABLE 2: Differences between children and adults in certain parameters affecting environmental exposures

Parameter	Newborn	Young Child	Older Child	Adult	Ref.
Surface area: body mass (m²/kg)	0.067	0.047	0.033	0.025	(18)
Respiratory ventilation rates	Infant			Adult	(18)
Respiratory volume (ml/kg per breath)	10			10	
Alveolar surface area (m²)	3			75	
Respiration rate (breaths/min)	40			15	
Respiratory minute ventilation rate[a]	133			2	
Mean drinking-water intake (ml/kg per day)	<1 year / 43.5	1-10 years / 35.5	11-19 years / 18.2	20-64 years / 19.9	(19)
Fruit consumption (g/kg per day (USA))	<1 year	3-5 years	12-19 years	40-69 years	(21)
Citrus fruits	1.9	2.6	1.1	0.9	
Other fruits (including apples)	12.9	5.8	1.1	1.3	
Apples	5.0	3.0	0.4	0.4	
Soil ingestion (mg/day)		2.5 years	6 years	Adult	(19)
Pica child		500			(5)
Outdoors		50	20	20[b]	
Indoors		60	2	0.4	
Differences in absorption of lead	42-53%	30-40%	18-24%	7-15%	(19)

[a] In ml/kg of body weight per m² lung surface area per minute.
[b] Gardening.

- Limited data are available on gene-environment interactions, an area identified as important for future research. Are those with certain genetic traits more likely to develop cancer or other health conditions associated with developmental exposures? Do these genetic traits impart greater vulnerability during particular developmental stages?
- The peripubertal/adolescent period is under-represented in studies of both exposure and outcomes in the current literature, despite the fact that many organ systems-especially the endocrine system-undergo significant development during this time.

More information is needed on windows of susceptibility in order to improve the risk assessment of potential environmental health threats to children, adolescents and adults. This requires increased interaction across different scientific disciplines. For example, animal laboratory data can help clinicians and epidemiologists identify areas of potential concern in humans. Epidemiological data can raise interest among laboratory researchers for developing mechanistic data and exploring agent-target interactions. Clinicians may contribute case reports that generate hypotheses for follow-up by other scientists. These interactions could help develop more sensitive methods for both clinicians and researchers for validation across species, and for enhancing the value of future studies on children's environmental health.

In summary, increased knowledge and information about children's windows of susceptibility to environmental agents will help to identify the particularly susceptible subgroups/ages and to plan for specific preventive actions. Increased dialogue among scientists from various disciplines will help to fill in the data gaps, improve measurement of exposures, and enhance the use of information to address and prevent the adverse effects of environmental threats on children's health.

REFERENCES

1.   Wilson JG. Embryological considerations in teratology. In: Wilson JG, Warkany J, eds. Teratology: principles and techniques. Chicago, The University of Chicago Press, 1965:256.
2.   Moore KL, Persaud TVN. The de~eloping human: clinically oriented embryology. Philadelphia, W.B. Saunders, 1913:98.

3.   Selevan SG, Lemasters GK. The dose-response fallacy in human reproductive stud-
     ies of toxic exposures. Journal of Occupational Medicine, 1987, 29: 451-454.
4.   Herbst AL, Scully RE, Robboy SJ. Prenatal diethylstilbestrol exposure and human
     genital tract abnormalities. National Cancer Institute Monographs, 1979, 51 :25-35·
5.   Horta BL et al. Low birthweight, preterm births and intrauterine growth retarda-
     tion in relation to maternal smoking. Paediatric and Perinatal Epidemiology,
     1997,11:140- 151.
6.   Gonzalez-Cossio T et al. Decrease in birth weight in relation to maternal bonelead
     burden. Pediatrics, 1997, 100:856- 862.
7.   Hu FB et al. Prevalence of asthma and wheezing in public schoolchildren: associa-
     tion with maternal smoking during pregnancy. Annals of Allergy, Asthma and Im-
     munology, 1997, 79:80- 84.
8.   Drews CD et al. The relationship between idiopathic mental retardation and mater-
     nal smoking during pregnancy. Pediatrics, 1996, 9:547-553·
9.   van Duijn CM et al. Risk factors for childhood acute non-lymphocytic leukemia: an
     association with maternal alcohol consumption during pregnancy? Cancer Epidemi-
     ology, Biomarkers and Prevention, 1994, 3:457-460.
10. Daniels JL, Olshan AF, Savitz DA. Pesticides and childhood cancers. Environmental
     Health Perspectives, 1997, 105:1068-1077.
11. Blanck HM et al. Age at menarche and tanner stage in girls exposed in utero and
     postnatally to polybrominated biphenyl. Epidemiology, 2000, 11:641- 647.
12. Selevan SG et al. Blood lead concentration and delayed puberty in girls. New Eng-
     land Journal of Medicine, 2003, 348:1527-1536.
13. Needleman H L et al. The long-term effects of exposure to low doses of lead in
     childhood. An 11-year follow-up report. New England Journal of Medicine, 1990,
     322:83-88.
14. Miller RW. Special susceptibility of the child to certain radiation-induced cancers.
     Environmental Health Perspectives, 103 (Su ppl. 6) :41-44.
15. Wadsworth ME, Kuh DJ. Childhood influences on adult health: a review of recent
     work from the British 1946 national birth cohort study, the M RC National Survey
     of Health and Development. Paediatric and Perinatal Epidemiology, 1997, 11:2-20.
16. Osmond C et al. Early growth and death from cardiovascular disease in women.
     British Medical Journal, 1993, 30T1519-1524.
17. Silvaggio T, Mattison DR. Comparative approach to toxicokinetics. In: Paul M, ed.
     Occupational and environmental reproductive hazards: a guide for clinicians. Balti-
     more, MD, Williams & Wilkins, 1993:25-36.
18. Snodgrass WR. Physiological and biochemical differences between children and
     adults as determinants of toxic response to environmental pollutants. In: Guze-
     lian PS, Henry CJ, Olin SS, eds. Similarities and differences between children and
     adults: implications for risk assessment. Washington, DC, I LS I Press, 1992: 35-42.
19. US Environmental Protection Agency. Exposure factors handbook. Vol 1: General
     factors. Washington, DC, 199T3-7 (EPA/600/P-95/002Fa).
20. Selevan SG, Kimmel CA, Mendola P, eds. Identifying critical windows of exposure
     for children's health. Environmental Health Perspectives, 2000, 108(Suppl 3):449-
     597.

21. US Environmental Protection Agency. Exposure factors handbook. Vol. 2: Food ingestion factors. Washington, DC, 199T9-13; 9-19 (EPA/600/P-95/002Fa).

CHAPTER 3

DEVELOPMENTAL MILESTONES IN CHILDREN'S ENVIRONMENTAL HEALTH

RUTH A. ETZEL

This month the children's environmental health (CEH) movement reaches the age of 21 years, an important milestone. The movement began in California in October 1989 with the launch of The Kids in the Environment Project, designed to train health professionals about environmental health and children. This project subsequently evolved into the Children's Environmental Health Network (http://www.cehn.org/), a national organization headquartered in Washington, DC. Pediatricians' interest in children's unique susceptibility to environmental hazards probably dates further back, at least to 1954, when fallout from a nuclear weapons test on Bikini Island in the South Pacific caused acute radiation burns among people living on neighboring islands. Two boys exposed to fallout when they were infants developed severe hypothyroidism and short stature in mid-childhood. Young children were more severely affected than adults; among 35 children < 15 years of age, 3 developed carcinoma of the thyroid, compared with 2 among 46 persons who were ≥ 15 years of age at exposure (Merke and Miller 1992). These cases highlighted children's

Used with permission from Environmental Health Perspectives. Etzel RA. Developmental Milestones in Children's Environmental Health. Environmental Health Perspectives, **118** (2010), http://dx.doi.org/10.1289/ehp.1002957

special vulnerability to ionizing radiation. Because of concerns about fall-out from weapons testing and fears of nuclear war, in 1957 the American Academy of Pediatrics established what is now known as the Committee on Environmental Health to address these issues. For 53 years this committee has been the world leader in advocacy and education about risks to children from environmental hazards, publishing the first book on pediatric environmental health in 1999 (Etzel 1999).

Despite these activities, modern medical and nursing school curricula in the United States have been slow to include information about the environment's impact on child health. In the early 1900s, the environment was already well understood to be an important contributor to health, and its importance was routinely taught to students. Think of Florence Nightingale's 6 Ds of disease: dirt, drink (clean drinking water), diet, damp, drafts, and drains (proper drainage and sewage systems). With the advent of high-tech medicine, however, these fundamentals began to receive short shrift in graduate education. At the same time, the house call—a home visit that allowed the physician to view the environment in which the patient lived—became increasingly rare, and doctors' visits began to take place in the confines of modern offices or hospitals, far removed from the day-to-day surroundings of the family. Over time, the environment became "invisible" to the medical practitioner.

To counteract this phenomenon, the CEH movement advocated for the development of specific research programs and training programs to introduce nursing and medical students to the concepts of children's vulnerability to environmental hazards. The CEH movement also encouraged academic and health organizations to study the impact of the environment on infant and child health. As a result, beginning in 1998, numerous Centers for Children's Environmental Health and Disease Prevention Research, funded by the National Institute of Environmental Health Sciences, were established in U.S. universities to study and prevent a range of childhood diseases and outcomes that can result from environmental exposures.

Nonetheless, in the early 2000s, physicians in the United States who were interested in careers in pediatric environmental health could find no formal fellowship training programs. In 2002, through the Academic Pediatric Association, the first formal fellowship training programs in pediatric environmental health were initiated. These 3-year training programs are designed to provide pediatricians with specific competencies to enable them to undertake environmental health research, teaching, and advocacy (Etzel et al. 2003). Fellowship training is available in many large U.S. cities including Boston, Massachusetts; New York, New York; Cincinnati, Ohio; Pittsburgh, Pennsylvania; San Francisco, California; and Seattle, Washington, as well as Vancouver, Canada.

Although the CEH movement began in the United States, within a decade it had become an international movement. In 1999 the World Health Organization (WHO) set up a Task Force for the Protection of Children's Environmental Health and began developing training materials on children's health and the environment for an international audience (Pronczuk de Garbino 2005; United Nations Environment Programme, United Nations Children's Fund, WHO 2002), including a Training Package for Health Care Providers, a set of peer-reviewed slide sets covering the major environmental issues for children (WHO 2010c). These slide sets (available free from WHO) have been used extensively to teach practicing physicians about pediatric environmental health issues not usually covered in the traditional medical school or postgraduate curriculum. In 2005, the International Pediatric Association, in collaboration with the WHO, launched the International Pediatric Environmental Health Leadership Institute to provide special training for doctors and nurses on children's health and the environment using the WHO Training Package for Health Care Providers (WHO 2010c). After attending the training course, participants who completed a community project, gave a presentation at their home hospital, and documented children's environmental diseases in their medical records were offered the opportunity to sit for an examination toward a special certificate in Children's Environmental Health. Diplomates of the Institute now provide training on CEH in many countries.

In addition to educating physicians, the WHO aimed to raise policy makers' awareness about children's health and the environment. To that end, the WHO has organized three large international conferences for this audience. In 2002 the 1st WHO International Conference on Environmental Threats to the Health of Children was held in Bangkok, Thailand. The conference focused on science-oriented issues, research needs, and capacity building while addressing the concrete needs for action and policies at the community, country, regional, and international levels. The 2nd WHO

International Conference on Children's Environmental Health was held in Buenos Aires, Argentina, in 2005. The 3rd WHO International Conference on Children's Health and the Environment was held in Busan, Republic of Korea, in 2009, resulting in a Global Plan of Action for Children's Health and the Environment (WHO 2010a).

This issue of Environmental Health Perspectives documents that substantial progress has been made in research to understand the role of the environment in the illnesses of childhood and adolescence. Consideration of illnesses traditionally associated with the environment, such as waterborne and foodborne diseases, has expanded to include study of natural toxins and toxic chemicals that derive from the rapid expansion of industry and technology. But much remains to be done to ensure that our burgeoning knowledge is translated into preventive actions for children.

Why has it been so difficult to move from knowing to doing? First, many of the decisions affecting children are made not by those in the health sector, but by our professional colleagues in the agriculture, education, energy, housing, mining, and transportation sectors. Just as "men are from Mars and women are from Venus," it seems as if professionals in each of these sectors are from different planets. Although we may speak the same language, we rarely have more than a cursory understanding of the forces that shape one another's decisions and other considerations. Because the CEH movement has focused on educating the health community, few efforts have been made to establish relationships with other economic sectors. To use a developmental analogy, we are still involved primarily in "parallel play" rather than team sports. Professionals in the health sciences may work alongside professionals in other sectors, but we are absorbed in our own activities and usually have little interaction outside them. Instead of sitting at the table with urban planners, housing specialists, and energy experts when health professionals are planning an approach to a child health problem such as asthma, we usually move forward to design a study, implement it, analyze the results, and then present it as a fait accompli to our colleagues in other economic sectors, and hope that they will find it useful.

This is not the ideal way to engage them. We medical professionals need to fully engage with other sectors as we launch our attempts to find solutions to child health problems. Major breakthroughs are likely to occur in protecting children from hazards in the environment only when we establish strong working relationships with those who haven't been trained as we have and who don't think as we do. One tool that helps different sectors to interact is Health Impact Assessment, promoted by the WHO (2010b) and by many countries including the United States. Health Impact Assessment helps decision makers make choices about alternatives and improvements to prevent disease/injury and to actively promote health. A recent White House Task Force on Childhood Obesity report recommends that communities consider integrating Health Impact Assessment into local decision-making processes before undertaking any major new development or planning initiative (White House Task Force on Childhood Obesity 2010).

Although pediatric environmental health, now 21 years old, can sit at the same table with other recognized pediatric subspecialities, and although some of the developmental milestones have been achieved, we have a ways to go in turning our considerable knowledge into action. Because the risk factors for environmentally related diseases reside in sectors beyond the direct control of public health, "parallel play" activities should give way to fully cooperative endeavors with other sectors.

REFERENCES

1. Etzel RA, ed. 1999. Handbook of Pediatric Environmental Health. Elk Grove Village, IL:American Academy of Pediatrics, Committee on Environmental Health.
2. Etzel RA, Crain EF, Gitterman BA, et al. 2003. Pediatric environmental health competencies for specialists. Ambul Pediatr 3:60–63.
3. Merke DP, Miller RW. 1992. Age differences in the effects of ionizing radiation. In: Similarities and Differences Between Children and Adults: Implications for Risk Assessment (Guzelian PS, Henry CJ, Olin SS, eds). Washington, DC:International Life Sciences Institute, 139–149.
4. Pronczuk de Garbino J, ed. 2005. Children's Health and the Environment: A Global Perspective. A Resource Manual for the Health Sector. Geneva:WHO. Available: http://whqlibdoc.who.int/publications/2005/9241562927_eng.pdf [accessed 13 September 2010].
5. United Nations Environment Programme, United Nations Children's Fund, WHO. 2002. Children in the New Millenium: Environmental Impact on Health. Geneva:United Nations Environment Programme, United Nations Children's Fund, WHO. Available: http://www.who.int/water_sanitation_health/hygiene/settings/millennium/en/index.html [accessed 13 September 2010]

6.   White House Task Force on Childhood Obesity. 2010. Solving the Problem of Childhood Obesity Within A Generation. Available: http://www.letsmove.gov/pdf/TaskForce_on_Childhood_Obesity_May2010_FullReport.pdf [accessed 13 September 2010].

7.   World Health Organization. 2010a. Global Plan of Action for Children's Health and the Environment. 2010. Available: http://www.who.int/cch/cchplanaction_10_15.pdf [accessed 13 September 2010].

8.   World Health Organization. 2010b. Health Impact Assessment: Promoting Health Across All Sectors of Activity. Available: http://www.who.int/hia/en/ [accessed 13 September 2010].

9.   World Health Organization. 2010c. Training Package for Health Care Providers. Available: http://www.who.int/ceh/capacity/trainpackage/en/index.html [accessed 13 September 2010].

PART II

EXPOSURE TO TOBACCO SMOKE

CHAPTER 4

PRENATAL TOBACCO SMOKE EXPOSURE IS ASSOCIATED WITH CHILDHOOD DNA CpG METHYLATION

CARRIE V. BRETON, KIMBERLY D. SIEGMUND, BONNIE R. JOUBERT, XINHUI WANG, WEILIANG QUI, VINCENT CAREY, WENCHE NYSTAD, SIRI E. HEBERG, CAROLE OBER, DAN NICOLAE, KATHLEEN C. BARNES, FERNANDO MARTINEZ, ANDY LIU, ROBERT LEMANSKE, ROBERT STRUNK, SCOTT WEISS, STEPHANIE LONDON, FRANK GILLILAND, AND BENJAMIN RABY ON BEHALF OF THE ASTHMA BRIDGE CONSORTIUM

4.1 INTRODUCTION

Smoking while pregnant is associated with a myriad of negative health outcomes both for the mother and for the fetus. [1] In utero tobacco smoke exposure (IUS) can damage the placental structure and function [2], is associated with changes in children's neurodevelopment and behavior [3]

as well as with impaired lung function and increased risk of developing asthma. [4], [5], [6] Moreover, IUS-related deficits in lung function are larger for children with asthma. [7].

One hypothesized mechanism through which IUS may act is by altering the epigenetic landscape within the developing fetus. Increasingly, scientific reports are linking IUS exposure to alterations in the fetal epigenome, including changes in DNA methylation in numerous genes and tissue types. [8], [9] To further investigate the association between IUS and epigenetics, we evaluated the association between IUS exposure and DNA methylation in the first phase of subjects participating in the Asthma BioRepository for Integrative Genomic Exploration (Asthma BRIDGE) study—a subset of asthmatic children originally from the Childhood Asthma Management Program (CAMP) trial. [10], [11] We designed a study aimed at investigating DNA methylation in promoter regions of genes as these are the most pertinent regulatory regions for gene expression. We interrogated 27,578 CpG loci using the Illumina HumanMethylation27 platform—a promoter-centric assay—measured in whole blood samples collected in the CAMP population and replicated our findings in two other populations: 1) the remainder of the Asthma BRIDGE project with Illumina HumanMethylation450 (HM450) data; and 2) the Norwegian Mother and Child Cohort Study (MoBa).

4.2 METHODS

4.2.1 STUDY POPULATION

Asthma BRIDGE is a multicenter initiative to develop a publicly accessible resource consisting of ~1,500 asthmatics and controls with comprehensive phenotype and genomic data. The first phase of the Asthma BRIDGE multicenter initiative consisted of asthmatic subjects that were originally recruited as part of CAMP, the details of which have been described elsewhere. [10], [11] Briefly, CAMP is a multicenter, randomized, double-masked clinical trial to compare the long-term effectiveness and safety of 3 inhaled treatments for asthma: budesonide, nedocromil, and placebo. Asthma was defined by the presence of asthma symptoms or the use of an inhaled bronchodilator at least twice per week or use of a daily asthma medication for the 6 months before the screening interview. All participants had increased airway responsiveness to methacholine (PC20<12.5 mg/mL) at study entry. Each parent or guardian signed a consent form and each participant 7 years of age and older signed an assent form approved by each clinical center's institutional review board. Prior to treatment randomization, DNA samples were collected from 968 of 1,041 willing CAMP participants, of which 572 were of self-reported Western European ancestry. Of these, sufficient DNA for methylation typing was available for 554 (97%). Mean age at DNA collection was 9 years.

4.2.2 REPLICATION WAS CONDUCTED IN THE REMAINDER OF ASTHMA BRIDGE SUBJECTS AND IN MOBA.

Asthma BRIDGE had 526 participants for whom whole blood samples with DNA methylation from the Illumina HM450 were available, of whom 332 were asthmatic. Subjects were recruited from five studies: Genomic Research on Asthma in the African Diaspora (GRAAD), Children's Health Study (CHS), Chicago Asthma Genetics (CAG), Childhood Asthma Research and Education (CARE), and Mexico City Childhood Asthma Study (MCCAS). Descriptions of these cohorts are provided in File S1. The primary objective of the Asthma BRIDGE initiative was to create a biorepository of cell lines and accompanying datasets for public access. All of the primary data analyzed for this study have been submitted to BioLINCC (https://biolincc.nhlbi.nih.gov), from where it will be made available to the public. The data is also submitted to dbGaP (http://www.ncbi.nlm.nih.gov/gap) within 6 months of publication.

The Norwegian Mother and Child Cohort Study (MoBa) consisted of 1062 participants in whom the association between maternal cotinine level and DNA methylation in cord blood was recently assessed using Illumina's Infinium HumanMethylation450 (HM450) BeadChip. [8] Maternal plasma cotinine concentrations in the MoBa samples were measured using liquid chromatography–tandem mass spectrometry. [12].

4.2.3 ETHICS STATEMENT

Written informed consent was provided by all study subjects. The institutional review boards (IRB) of the Brigham and Women's Hospital, and each of the participating CAMP study centers, approved these protocols. The MoBa study was approved Norwegian Data Inspectorate and the Regional Ethics Committee for Medical Research and the NIEHS Institutional Review Board.

4.2.4 MATERNAL SMOKING EXPOSURE

Information regarding IUS exposure was obtained during the baseline medical interview for CAMP and Asthma BRIDGE subjects. Children were considered to have IUS exposure if respondents answered affirmatively to the question, "Did this child's mother smoke while she was pregnant with this child?" Information about current environmental tobacco smoke exposure was obtained during follow-up interviews with the question, "Do any caretakers of the child currently smoke cigarettes?" In MoBa, cotinine concentrations were measured in maternal samples collected at about 18 weeks of pregnancy using liquid chromatography–tandem mass spectrometry. [12], [13]

4.2.5 DNA METHYLATION

Laboratory personnel performing DNA methylation analysis were blinded to study subject information. DNA was extracted whole blood cells using the QiaAmp DNA blood kit (Qiagen Inc, Valencia, CA) and stored at −80 degrees Celcius. Two micrograms of genomic DNA from each sample were treated with bisulfite using the EZ-96 DNA Methylation Kit (Zymo Research, Irvine, CA, USA), according to the manufacturer's recommended protocol and eluted in 18 ul. The results of the HM27 assay were compiled for each locus as previously described and were reported as beta (β) values. [14] Details of quality control procedures, dilution control series, and calculation of β values are provided in File S1.

CpG loci on the HM27 array were removed from analyses if they were on the X and Y chromosomes, or if they contained SNPs, deletions, repeats, or if they mapped to multiple places in the genome. A Normal exponential background correction was first applied to the raw intensities at the array level to reduce background noise. [15] We then normalized each sample's methylation values to have the same quantiles to address sample to sample variability. [16] Lastly, we applied COMBAT, a mixed effects model that accounted for exposure, to reduce probe level variation. [17] The density plots after each step are shown in Figure S1 (in File S1), and illustrate the improvement acquired from each correction applied.

Samples from 554 CAMP participants were included for initial analysis of DNA methylation. Six samples were excluded for having methylation call rates <95% and an additional 17 were excluded because their sample mean methylation across all probes was greater than 2 standard deviations from the overall mean of all samples on the plate. Two additional samples were removed for discrepancy of sex determination when comparing questionnaire response data to X chromosome methylation patterns. One additional sample was removed for missing IUS exposure and another was removed before COMBAT correction for being the only sample on a single chip. This left 527 samples for analysis.

4.2.6 MRNA EXPRESSION

Paired samples of isolated mRNA and DNA collected at the same point in time were available for the Asthma BRIDGE replication dataset only. We collected 5 cc of whole blood in RNA PaxGene tubes. Samples were shipped to the data coordinating center monthly for RNA extraction using the PAXgene Blood RNA Kit according to manufacturer's protocol (PreAnalytix). Expression profiles were generated using the Illumina Human HT-12 v4 arrays, according to manufacturers' protocol (Illumina, San Diego CA). Preprocessing was performed using quantile normalization with the lumi Bioconductor package. [18] Evidence for association between CpG methylation and target gene expression was assessed using linear regression, including adjustment for age, gender, and clinical center.

4.2.7 STATISTICAL ANALYSES

Descriptive analyses were performed to examine the distribution of subject characteristics. Density plots of DNA methylation values were created and evaluated for quality control. Outlier DNA methylation values were identified as values that were either greater than the median+3*IQR or less than the median- 3*IQR and were removed from analyses.

To investigate the association between IUS exposure and percent DNA methylation, we fitted beta regression models adjusted for age, sex, clinic and cell type. The following cell types were estimated using the method of Houseman et al [19]: B-lymphocytes, granulocytes, monocytes, natural killer cells, CD4+ T-lymphocytes, and CD8+ T-lymphocytes. Beta regression was used to address the non-normal distribution of DNA methylation values, which are bounded by 0 and 1 and in many cases heavily skewed toward one end or the other. [20] The beta regression model is as follows: Let $y_1,...,y_n$ be a random sample of DNA methylation beta values for a single feature on the HM27 assay. We assume $y_i \sim Beta(\mu_i, \varphi)$, $i = 1,...,n$, with $E(y) = \mu$ and $Var(y) = \mu(1-\mu)/(1+\varphi)$, the parameter φ measuring precision. The regression model is defined as $logit(\mu_i) = a + a_1 X_{i1} + ... + a_k X_{ik}$, with X_1 measuring IUS exposure and $X_2,...,X_k$ measuring age, sex, clinic, and estimated cell type fractions for B-lymphocytes, granulocytes, monocytes, natural killer cells, CD4+ T-lymphocytes, and CD8+ T-lymphocytes. The precision parameter is assumed constant for all observations. All regressions were run in R using the betareg package (version R2.15.3).

Beta regression requires the data to be between zero and one, therefore a shrinkage method was applied to force the zeros to be positive as follows: 0.999999*(meth value-0.5) +0.5. Because beta regression is modeled on the logit scale, we also present estimates from linear regression models to provide interpretation of a % difference in methylation and compare effect estimates to those from our replication population. All regression analyses were adjusted for multiple testing at a false discovery rate (FDR) of 0.05, using the method of Benjamini and Hochberg. [21]

To investigation the association between methylation or IUS and expression, linear regression models were fit in which expression values were transformed on the log2 scale. Sensitivity analyses were conducted in regression models of our top 19 loci to evaluate potential confounding by the following covariates: income, education, parental history of asthma, asthma severity, maternal or paternal smoking in childhood. To replicate results, we ran identical models in the 526 participants in Asthma BRIDGE, as well as within asthmatics only. We also compared our results to those from a study of 1062 participants from the Norwegian Mother and Child Cohort Study. [8] Joubert et al examined the association between maternal plasma cotinine (active smoking defined as >56.8 nmol/L) and cytosine methylation in cord blood using robust linear regression on the log-ratio of the methylation beta value obtained from the Illumina HM450 array, adjusting for several covariates.

All tests assumed a two-sided alternative hypothesis, a 0.05 significance level, and were conducted using the R programming language, version R2.15.3.

4.3 RESULTS

The HM27 assay demonstrated high reproducibility, with an average spearman correlation coefficient of 0.96 across all replicate control samples (i.e. PBL and dilution controls). Additional analyses of accuracy, assessed by calculating bias within the dilution controls, suggested minimal bias, with values of 0.15, 0.07, −0.03, and −0.11 for 10%, 35%, 60% and 85% dilution controls, respectively.

Of the 527 children in the study, 65 (12.3%) were exposed to maternal smoking during pregnancy. Age and sex did not vary between exposed and unexposed children. However, unexposed children were more likely to have a maternal (26%) or paternal (21%) history of asthma compared to exposed children (16%). Mothers who smoked during pregnancy also had significantly lower education and income levels than mothers who did not smoke (Table 1).

After application of background correction and normalization procedures, we observed a total of 26 loci that were statistically significantly associated with IUS at a FDR <0.05. However, 7 of these loci were removed from further consideration because their mean methylation levels were less than 3% or greater than 97% and thus largely invariant. The remaining 19 loci are shown in Table 2.

TABLE 1: Descriptive characteristics of the CAMP study population (n = 527).

Characteristics	level	Overall Count	Overall %	No (N = 462) Count	No (N = 462) %	Yes (N = 65) Count	Yes (N = 65) %	p-values*
In utero tobacco smoke exposure		65	12.3					
Male		320	60.7	280	60.6	40	61.5	0.89
Maternal history of asthma		129	25.0	119	26.2	10	16.1	0.09
Paternal history of asthma		103	20.6	94	21.3	9	15.8	0.34
Maternal education	≤ high school	87	16.5	60	13.0	27	41.5	<0.0001
	some college	203	38.6	175	38.0	28	43.1	
	college degree	236	44.9	226	49.0	10	15.4	
Clinic site	Albuquerque	45	8.5	37	8.0	8	12.3	0.15
	Baltimore	79	15.0	64	13.9	15	23.1	
	Boston	41	7.8	36	7.8	5	7.7	
	Denver	68	12.9	61	13.2	7	10.8	
	San Diego	53	10.1	50	10.8	3	4.6	
	Seattle	86	16.3	80	17.3	6	9.2	
	St. Louis	91	17.3	81	17.5	10	15.4	
	Toronto	64	12.1	53	11.5	11	16.9	
Age (mean (SD))		8.7	2.1	8.7	2.1	8.7	2.1	0.92

The column group "Exposure to Maternal Smoking *in utero*" spans the No (N = 462), Yes (N = 65), and p-values* columns.

p-values are calculated using chi-sq test.

We evaluated the 19 CpGs in two replication datasets (Tables S1, S2 in File S1). In the Asthma BRIDGE replication population, three of the 19 loci were not available. Of the remaining 16, four loci had nominal p-values of less than 0.05: *XPNPEP1* (p = 0.01), *PPEF2* (p = 0.003), *FR-MD4A* (p = 0.01), *C11orf52* (p = 0.04) (Table 3, Figure 1). *XPNPEP1* and *PPEF2* remained statistically significant after FDR adjustment for multiple testing. Moreover, the results were consistent if we further restricted this population to asthmatics only, to be more directly comparable to the CAMP population, suggesting that the maternal smoking effects on these genes are not restricted to asthmatic children. In MoBa, one locus was not found in that dataset and four loci had nominal p-values of less than 0.05:

FRMD4A (p = 0.0009), *C11orf52* (p = 0.001), *SMPD3* (p = 0.02), and *CRYGN* (p = 0.05) (Table 3). *FRMD4A* and *C11orf52* remained significant after further adjustment for multiple testing. Interestingly, for some of these genes multiple CpG loci within the gene exhibited similar effect sizes and were located near the transcription start sites for *FRMD4A*, *CRYGN*, and *PPEF2* (Tables S3, S4 in File S1).

TABLE 2: CpG Loci significantly associated with IUS exposure in CAMP asthmatics (N = 527).

Probe ID	Symbol	Chr	Mean methylation level	β*	Difference in methylation**	FDR p-value
cg05697249	C11orf52	11	0.66	0.09	0.02	0.001
cg14724265	PPEF2	4	0.76	0.17	0.03	0.01
cg25464840	FRMD4A	10	0.75	0.13	0.02	0.01
cg20588045	PSDH15	10	0.28	0.10	0.02	0.01
cg20555507	TRPM3	9	0.08	0.13	0.01	0.03
cg16184943	ZNF280B	22	0.34	−0.09	−0.02	0.03
cg09352789	XPNPEP1	10	0.29	−0.06	−0.01	0.03
cg07499072	FST	5	0.04	0.11	0.004	0.03
cg22830895	CRYGN	7	0.43	0.11	0.03	0.03
cg13473383	ZDHHC5	11	0.09	−0.08	−0.01	0.03
cg10556064	SMPD3	16	0.58	0.10	0.02	0.03
cg04112019	IGF2AS	11	0.12	0.07	0.01	0.03
cg00169548	BAZ1A	14	0.48	−0.10	−0.02	0.03
cg10493739	TMEM38B	9	0.06	−0.07	−0.004	0.03
cg01058368	CDH10	5	0.82	0.12	0.02	0.04
cg09143663	BACH1	21	0.72	0.08	0.02	0.04
cg20773127	ENPEP	4	0.58	0.10	0.02	0.05
cg24956866	CALD1	7	0.58	0.10	0.02	0.05
cg14580737	RFXANK	19	0.80	−0.10	−0.02	0.05

*coefficient from beta regression adjusted for age, sex, and clinic and cell type.
**coefficient from linear regression model illustrating difference in methylation level comparing IUS exposed to unexposed.

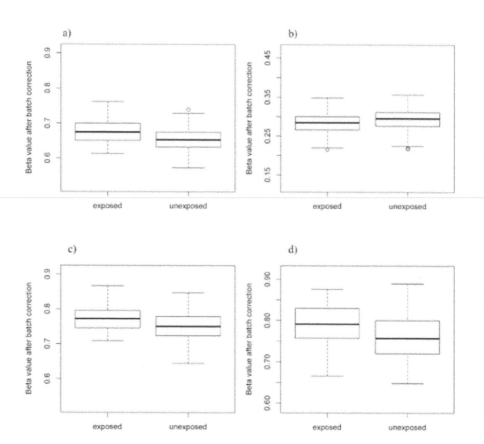

FIGURE 1: Boxplots showing the distribution of % CpG methylation after normalization and batch correction using COMBAT by IUS exposure for A) *C11orf52* (cg05697249), B) *XPNPEP1* (cg09352789), C) *FRMD4A* (cg25464840), and D) *PPEF2* (cg14724265) in 526 Asthma BRIDGE samples.

TABLE 3: Replication results from beta regression and linear regression models in Asthma BRIDGE and MoBa for the 19 CpG loci identified in CAMP.

Probe ID	Symbol	Chr	ABRIDGE (N = 526)		ABRIDGE Asthma only (N = 332)		Difference in methylation**		
			β*	FDR p-value	β*	FDR p-value	CAMP	ABRIDGE	MoBa
cg09352789	*XPNPEP1*	10	−0.07	0.04	−0.08	0.12	−0.01	−0.01***	−0.004
cg14724265	*PPEF2*	4	0.11	0.04	0.12	0.12	0.03	0.02***	−0.009
cg25464840	*FRMD4A*	10	0.08	0.07	0.08	0.18	0.02	0.02	0.03***
cg05697249	*C11orf52*	11	0.06	0.14	0.06	0.22	0.02	0.01	0.02***
cg16184943	ZNF280B	22	−0.05	0.20	−0.06	0.22	−0.02	−0.01	0
cg04112019	IGF2AS	11	−0.05	0.29	−0.07	0.22	0.01	−0.003	−0.01
cg13473383	ZDHHC5	11	−0.04	0.29	−0.06	0.22	−0.01	−0.01	−0.001
cg20773127	ENPEP	4	0.04	0.29	0.03	0.63	0.02	0.01	0.003
cg14580737	RFXANK	19	−0.05	0.37	−0.06	0.37	−0.02	−0.004	0.004
cg22830895	*CRYGN*	7	−0.02	0.84	−0.02	0.69	0.03	−0.01	0.02
cg01058368	CDH10	5	0.0004	0.99	−0.01	0.90	0.02	0	0.01
cg07499072	FST	5	−0.002	0.99	−0.01	0.90	0.00	0	−0.003
cg09143663	BACH1	21	0.01	0.99	0.05	0.48	0.02	0.001	0.01
cg10556064	*SMPD3*	16	−0.01	0.99	−0.002	0.96	0.02	−0.001	0.02
cg20588045	PCDH15	10	−0.01	0.99	−0.01	0.90	0.02	−0.002	−0.002
cg24956866	CALD1	7	0.001	0.99	0.03	0.63	0.02	0.002	0.01
cg00169548	BAZ1A	14	n/a	n/a	n/a	n/a	−0.02	n/a	n/a
cg10493739	TMEM38B	9	n/a	n/a	n/a	n/a	0.00	n/a	−0.001
cg20555507	TRPM3	9	n/a	n/a	n/a	n/a	0.01	n/a	0.01

*coefficient from beta regression adjusted for age, sex, and clinic and cell type.
**coefficient from linear regression model illustrated difference in methylation level comparing IUS exposed to unexposed.
***FDR corrected p-value <0.05

In the initial screening analysis, the model was adjusted only for sex, age clinic, and cell type. We conducted additional analyses to evaluate whether the association was confounded by other potentially relevant factors. No evidence for confounding was noted for income, education, parental history of asthma, asthma severity, or exposure to paternal smoking in childhood (data not shown). As expected, report of exposure to maternal smoking during childhood was strongly associated with in utero smoke exposure, reported in 57 (88%) of 65 in utero exposed children.

Evidence for collinear association by maternal smoking in childhood was present for some but not all of the loci, in some cases strengthening the association and in some cases decreasing the association for prenatal tobacco smoke exposure (Table S5 in File S1). For example, adjustment for childhood smoke exposure had little to no effect on the estimate for prenatal smoke exposure for *C11orf52* and *PPEF2*. However, adjustment for childhood smoke exposure attenuated the effect estimates for *FRMD4A* and *XPNPEP1*.

The direction and magnitude of effect estimates for IUS exposure was consistent in all datasets for all loci. Interestingly, *FRMD4A* and *C11orf52* showed consistently strong effect estimates across all three diverse populations, regardless of whether the children were asthmatic, and regardless of the timing of sample collection. In fact, the use of cord blood in the MoBa replication dataset lends support to the conclusion that the effects on DNA methylation in *FRMD4A* and *C11orf52* are due to prenatal rather than childhood smoke exposure.

Lastly, in the Asthma BRIDGE replication population, we evaluated the relation between DNA methylation and expression for *XPNPEP1*, *PPEF2*, and *FRMD4* as well as the association between IUS and expression. *C11orf52* could not be evaluated as there was no associated mRNA transcript. Though expression profiles in whole blood were available for 526 subjects, we did not observe significant associations between methylation and expression for the three tested genes. However, expression of two of these genes (*PPEF2* and *XPNPEP1*) was significantly decreased in subjects with IUS exposure compared to those without IUS exposure ($\beta = -0.02$, $p = 0.02$ for *PPEF2*; and $\beta = -0.05$, $p = 0.01$ for *XPNPEP1*), demonstrating that exposure was associated both with changes in both methylation and expression in the same sample.

4.4 DISCUSSION

In a comprehensive investigation of IUS exposure and DNA methylation in the offspring, we identified 19 CpG loci in whole blood significantly associated with IUS in asthmatic children, two of which (*FRMD4*A and

C11orf52) were replicated in two additional independent populations and four others that warrant potential further investigation.

*FRMD4*A encodes a scaffolding protein involved in activation of Arf6, which in turn, plays a role in membrane trafficking, junctional remodeling and epithelial polarization. [22] Our observation of increased DNA methylation in *FRMD4A* in whole blood samples from children exposed to maternal smoking is consistent with an experimental study showing hypermethylated *FRMD4A* in MCF-7 cells treated with benzo(a)pyrene (BaP), a common polycyclic aromatic hydrocarbon that is a constituent of cigarette smoke. [23] Moreover, *FRMD4A* was recently identified from a GWAS study in Asian populations and validated in European and American populations as a gene associated with nicotine dependence. [24] IUS exposure is also associated with nicotine dependence in offspring. [25], [26] These observations raise the intriguing possibility that cigarette-smoke-induced epigenetic modification of the *FRMD4A* gene plays a role in conferring increased risk of nicotine dependence in offspring of mothers who smoke during pregnancy. Beyond nicotine dependence, altered DNA methylation of *FRMD4A* may have the potential to affect other downstream health outcomes, since the gene has recently been implicated as a risk factor for both Alzheimer's disease and squamous cell carcinoma. [27], [28]

Very little is currently known about the other replicated locus within *C11orf52*. *C11orf52* describes an uncharacterized protein that is expressed in the lung and which has been associated with increased phosphorylation in non-small cell lung cancer tumor compared to normal tissue samples. [29] However, the *C11orf52* locus also overlaps with heat shock 27kDa protein 2 (*HSPB2*). *HSPB2* is a stress-inducible small heat shock protein that plays a role in airway smooth muscle (ASM) cell remodeling. *HSPB2* undergoes rapid phosphorylation in the ASM cell thereby stabilizing the cytoskeletal scaffolding and decreasing the rate of microstructural reorganization. [30] ASM remodeling plays a role in airway hyperresponsiveness and is a cardinal feature of asthma. [31] Moreover, particulate matter exposure was shown to stimulate reactive oxygen species generation in human lung vascular endothelium, resulting in increased *HSPB2* phosphorylation and a marked disruption in endothelial cell barrier function via cytoskeletal rearrangement, key elements for lung homeostasis. [32]

Our observation of an IUS-induced change in CpG methylation level in a locus within *HSPB2* in asthmatic children raises the interesting question of whether these changes directly affect the function of or phosphorylation of *HSPB2* and could thus play a role in airway remodeling.

XPNPEP1 and *PPEF2* are two genes that may also be of interest as they were significant in both the CAMP and Asthma BRIDGE populations. However, little is known about these genes. *XPNPEP1* located on chromosome 10, encodes the human soluble aminopeptidase P (APP) which is an aminoacylprolyl hydrolase. [33] APP is a key enzyme in the renin-angiotensin and kininogen-kinin hormonal systems and has been associated with oral contraceptive use. [34] Specifically APP degrades bradykinin, a blood pressure regulator peptide, and has been linked to myocardial infarction. [35] *PPEF2* is a protein phosphatase with EF-hand domain whose function is not well understood. *PPEF2* has also been implicated in dendritic cell development in a murine model [36] and with schizophrenia in a small population. [37] *PPEF2* expression correlates with stress protective responses, cell survival, growth and proliferation and is a negative regulator of apoptosis signal regulating kinase-1 (ASK1), important in cancer, cardiovascular, and neurodegenerative diseases. [38], [39] Our observation that IUS was associated with decreased expression in *PPEF2* may not be surprising, given *PPEF2*'s purported role in stress response. However, we provide some of the first evidence for an association between IUS exposure and expression and CpG methylation in both *PPEF2* and XPNPEP; thus, a much greater understanding of the biological roles of these genes is necessary before we can understand the health implications of IUS on them.

SMPD3 and *CRYGN* are additional genes for which DNA methylation was significantly associated with IUS exposure in both the CAMP and MoBA, but did not meet the stringent criteria for replication after adjustment for multiple testing. *SMPD3* is the primary regulator of ceramide biosynthesis, and plays a pivotal role in the control of late embryonic and postnatal development such that defects in the gene lead to dwarfism and pituitary hormone deficiency. [40], [41] *SMPD3* has also been implicated in macrophage differentiation and leukemia. [42] Given *SMPD3*'s crucial role in development and its expression in neurons in the central nervous system, our observed association with IUS raises intriguing questions about the biological mechanisms underlying smoking's affect on prenatal growth. Lastly, *CRYGN* encodes a crystallin, one of the main structural proteins in the eye. [43] Very little research on *CRYGN* currently exists, thus the implications of our association with IUS with regard to developmental biology are largely unknown.

In this study, the associations between IUS and DNA methylation of several genes were observed using questionnaire-based recall of maternal smoking as well as measured cotinine levels in cord blood. While questionnaire-based recall of maternal smoking history is susceptible to recall bias, the positive findings reported in a previous analysis in the CAMP cohort supports the validity of the retrospective history. [44] DNA methylation was measured both at birth in cord blood (MoBa) and in whole blood at age 9 (CAMP) and in adults (Asthma BRIDGE). Our robust results suggest the IUS effects on DNA methylation may persist over years at least for some genes. An additional strength of our use of cord blood for replication is that we remove potential confounding effects of childhood maternal or paternal smoking exposure, and can more confidently conclude the observed effects are due to in utero exposure. On the other hand, because our primary and replication studies were performed using whole blood collected at different time points, we may have missed transient yet important effects on DNA methylation.

DNA methylation was measured in whole blood, composed of a variable mixture of circulating cell types. In theory, a shift in cell populations caused by IUS, rather than differences in DNA methylation, could explain our observed results. In order for this to be true, however, IUS would have to alter the proportions of cell populations not only at birth but 9 or more years later. To shed further light on this possibility, we estimated 6 cell types using the method of Houseman et al [19] and included these estimated cell types as covariates in the regression model. Additionally, we tested whether IUS exposure was associated with cell type and found no association. Given these results, it is unlikely that differences in cell types are accounting for our observed associations. Because we evaluated DNA methylation in whole blood, we do not know whether these same observations would be apparent in other tissue or cell types of interest.

DNA methylation in our loci was not highly correlated with expression in whole blood. However, IUS exposure was significantly associated with decreased expression levels in *PPEF2* and *XPNPEP1*. One reason for the lack of direct correlation between methylation and expression may be because the methylated loci for *PPEF2* and *XPNPEP1* lie within gene bodies, and not the promoters typically associated with expression levels, [45] and thus may be only one piece in a complex regulatory network. An additional reason may be that these genes are functionally relevant only in utero or shortly after birth, during the time at which maternal smoking exposure occurred. While the methylation marks have endured and represent a biomarker of past exposure, the functionality of the gene may change with developmental time period and tissue.

Lastly, we acknowledge that the use of the HM27 assay is not an epigenome-wide interrogation of DNA methylation. Nevertheless, the HM27 assay served our hypothesis well, since we hypothesized IUS exposure would be associated with CpG methylation in promoter regions of genes that regulate expression.

In summary, our results showing an increase in DNA methylation level at two loci in *FRMD4A* and Cllorf52 add to the growing evidence that IUS exposure in humans has epigenetic consequences for the offspring, some of which may persist throughout childhood. Though the long term health implications of these loci are currently unknown and require further investigation, their implication in nicotine dependence and lung homeostasis suggests their potential contribution to the development of disease in later life.

REFERENCES

1.   Finkelstein JB (2006) Surgeon General's report heralds turning tide against tobacco, smoking. J Natl Cancer Inst 98: 1360–1362. doi: 10.1093/jnci/djj427
2.   Jauniaux E, Burton GJ (2007) Morphological and biological effects of maternal exposure to tobacco smoke on the feto-placental unit. Early Hum Dev 83: 699–706. doi: 10.1016/j.earlhumdev.2007.07.016
3.   Eskenazi B, Castorina R (1999) Association of prenatal maternal or postnatal child environmental tobacco smoke exposure and neurodevelopmental and behavioral problems in children. Environ Health Perspect 107: 991–1000. doi: 10.1289/ehp.99107991
4.   Burke H, Leonardi-Bee J, Hashim A, Pine-Abata H, Chen Y, et al. (2012) Prenatal and passive smoke exposure and incidence of asthma and wheeze: systematic review and meta-analysis. Pediatrics 129: 735–744. doi: 10.1542/peds.2011-2196

5. Wang L, Pinkerton KE (2008) Detrimental effects of tobacco smoke exposure during development on postnatal lung function and asthma. Birth Defects Res C Embryo Today 84: 54–60. doi: 10.1002/bdrc.20114

6. Gilliland FD, Li YF, Peters JM (2001) Effects of maternal smoking during pregnancy and environmental tobacco smoke on asthma and wheezing in children. Am J Respir Crit Care Med 163: 429–436. doi: 10.1164/ajrccm.163.2.2006009

7. Li YF, Gilliland FD, Berhane K, McConnell R, Gauderman WJ, et al. (2000) Effects of in utero and environmental tobacco smoke exposure on lung function in boys and girls with and without asthma. Am J Respir Crit Care Med 162: 2097–2104. doi: 10.1164/ajrccm.162.6.2004178

8. Joubert BR, Haberg SE, Nilsen RM, Wang X, Vollset SE, et al. (2012) 450K Epigenome-Wide Scan Identifies Differential DNA Methylation in Newborns Related to Maternal Smoking during Pregnancy. Environ Health Perspect 120: 1425–1431. doi: 10.1289/ehp.1205412

9. Suter MA, Anders A, Aagaard KM (2012) Maternal Smoking as a Model for Environmental Epigenetic Changes Affecting Birth Weight and Fetal Programming. Mol Hum Reprod.

10. The Childhood Asthma Management Program (CAMP): design, rationale, and methods. Childhood Asthma Management Program Research Group. Control Clin Trials 20: 91–120. doi: 10.1016/s0197-2456(98)00044-0

11. Long-term effects of budesonide or nedocromil in children with asthma. The Childhood Asthma Management Program Research Group. N Engl J Med 343: 1054–1063. doi: 10.1056/nejm200010123431501

12. Midttun O, Hustad S, Ueland PM (2009) Quantitative profiling of biomarkers related to B-vitamin status, tryptophan metabolism and inflammation in human plasma by liquid chromatography/tandem mass spectrometry. Rapid Commun Mass Spectrom 23: 1371–1379. doi: 10.1002/rcm.4013

13. Kvalvik LG, Nilsen RM, Skjaerven R, Vollset SE, Midttun O, et al. (2012) Self-reported smoking status and plasma cotinine concentrations among pregnant women in the Norwegian Mother and Child Cohort Study. Pediatr Res 72: 101–107. doi: 10.1038/pr.2012.36

14. Noushmehr H, Weisenberger DJ, Diefes K, Phillips HS, Pujara K, et al. (2010) Identification of a CpG island methylator phenotype that defines a distinct subgroup of glioma. Cancer Cell 17: 510–522. doi: 10.1016/j.ccr.2010.03.017

15. Triche TJ, Weisenberger DJ, Van Den Berg D, Laird PW, Siegmund KD (2013) Low-level Processing of Illumina Infinium DNA Methylation BeadArrays. Nucleic Acids Res (in press).

16. Bolstad BM, Irizarry RA, Astrand M, Speed TP (2003) A comparison of normalization methods for high density oligonucleotide array data based on variance and bias. Bioinformatics 19: 185–193. doi: 10.1093/bioinformatics/19.2.185

17. Johnson WE, Li C, Rabinovic A (2007) Adjusting batch effects in microarray expression data using empirical Bayes methods. Biostatistics 8: 118–127. doi: 10.1093/biostatistics/kxj037

18. Du P, Kibbe WA, Lin SM (2008) lumi: a pipeline for processing Illumina microarray. Bioinformatics 24: 1547–1548. doi: 10.1093/bioinformatics/btn224

19. Houseman EA, Accomando WP, Koestler DC, Christensen BC, Marsit CJ, et al. (2012) DNA methylation arrays as surrogate measures of cell mixture distribution. BMC Bioinformatics 13: 86. doi: 10.1186/1471-2105-13-86
20. Ferrari SLP, Cribari-Neto F (2004) Beta regression for modelling rates and proportions. Journal of Applied Statistics 31: 799–815. doi: 10.1080/0266476042000214501
21. Hochberg Y, Benjamini Y (1990) More powerful procedures for multiple significance testing. Stat Med 9: 811–818. doi: 10.1002/sim.4780090710
22. Ikenouchi J, Umeda M (2010) *FRMD4A* regulates epithelial polarity by connecting Arf6 activation with the PAR complex. Proc Natl Acad Sci U S A 107: 748–753. doi: 10.1073/pnas.0908423107
23. Sadikovic B, Andrews J, Rodenhiser DI (2007) DNA methylation analysis using CpG microarrays is impaired in benzopyrene exposed cells. Toxicol Appl Pharmacol 225: 300–309. doi: 10.1016/j.taap.2007.08.013
24. Yoon D, Kim YJ, Cui WY, Van der Vaart A, Cho YS, et al. (2012) Large-scale genome-wide association study of Asian population reveals genetic factors in *FRMD4A* and other loci influencing smoking initiation and nicotine dependence. Hum Genet 131: 1009–1021. doi: 10.1007/s00439-011-1102-x
25. O'Callaghan FV, Al Mamun A, O'Callaghan M, Alati R, Najman JM, et al. (2009) Maternal smoking during pregnancy predicts nicotine disorder (dependence or withdrawal) in young adults - a birth cohort study. Aust N Z J Public Health 33: 371–377. doi: 10.1111/j.1753-6405.2009.00410.x
26. Rydell M, Cnattingius S, Granath F, Magnusson C, Galanti MR (2012) Prenatal exposure to tobacco and future nicotine dependence: population-based cohort study. Br J Psychiatry 200: 202–209. doi: 10.1192/bjp.bp.111.100123
27. Lambert JC, Grenier-Boley B, Harold D, Zelenika D, Chouraki V, et al.. (2012) Genome-wide haplotype association study identifies the *FRMD4A* gene as a risk locus for Alzheimer's disease. Mol Psychiatry.
28. Goldie SJ, Mulder KW, Tan DW, Lyons SK, Sims AH, et al. (2012) *FRMD4A* upregulation in human squamous cell carcinoma promotes tumor growth and metastasis and is associated with poor prognosis. Cancer Res 72: 3424–3436. doi: 10.1158/0008-5472.can-12-0423
29. Wu CJ, Cai T, Rikova K, Merberg D, Kasif S, et al. (2009) A predictive phosphorylation signature of lung cancer. PLoS One 4: e7994. doi: 10.1371/journal.pone.0007994
30. An SS, Fabry B, Mellema M, Bursac P, Gerthoffer WT, et al. (2004) Role of heat shock protein 27 in cytoskeletal remodeling of the airway smooth muscle cell. J Appl Physiol 96: 1701–1713. doi: 10.1152/japplphysiol.01129.2003
31. King GG, Pare PD, Seow CY (1999) The mechanics of exaggerated airway narrowing in asthma: the role of smooth muscle. Respir Physiol 118: 1–13. doi: 10.1016/s0034-5687(99)00076-6
32. Wang T, Chiang ET, Moreno-Vinasco L, Lang GD, Pendyala S, et al. (2010) Particulate matter disrupts human lung endothelial barrier integrity via ROS- and p38 MAPK-dependent pathways. Am J Respir Cell Mol Biol 42: 442–449. doi: 10.1165/rcmb.2008-0402oc
33. Sprinkle TJ, Caldwell C, Ryan JW (2000) Cloning, chromosomal sublocalization of the human soluble aminopeptidase P gene (*XPNPEP1*) to 10q25.3 and conservation

of the putative proton shuttle and metal ligand binding sites with XPNPEP2. Arch Biochem Biophys 378: 51–56. doi: 10.1006/abbi.2000.1792

34. Cilia La Corte AL, Carter AM, Turner AJ, Grant PJ, Hooper NM (2008) The bradykinin-degrading aminopeptidase P is increased in women taking the oral contraceptive pill. J Renin Angiotensin Aldosterone Syst 9: 221–225. doi: 10.1177/1470320308096405

35. Li X, Lou Z, Zhou W, Ma M, Cao Y, et al. (2008) Structure of human cytosolic X-prolyl aminopeptidase: a double Mn(II)-dependent dimeric enzyme with a novel three-domain subunit. J Biol Chem 283: 22858–22866. doi: 10.1074/jbc.m710274200

36. Edelmann SL, Nelson PJ, Brocker T (2011) Comparative promoter analysis in vivo: identification of a dendritic cell-specific promoter module. Blood 118: e40–49. doi: 10.1182/blood-2011-03-342261

37. Timms AE, Dorschner MO, Wechsler J, Choi KY, Kirkwood R, et al. (2013) Support for the N-methyl-D-aspartate receptor hypofunction hypothesis of schizophrenia from exome sequencing in multiplex families. JAMA Psychiatry 70: 582–590. doi: 10.1001/jamapsychiatry.2013.1195

38. Kutuzov MA, Bennett N, Andreeva AV (2010) Protein phosphatase with EF-hand domains 2 (*PPEF2*) is a potent negative regulator of apoptosis signal regulating kinase-1 (ASK1). Int J Biochem Cell Biol 42: 1816–1822. doi: 10.1016/j.biocel.2010.07.014

39. Buchser WJ, Slepak TI, Gutierrez-Arenas O, Bixby JL, Lemmon VP (2010) Kinase/phosphatase overexpression reveals pathways regulating hippocampal neuron morphology. Mol Syst Biol 6: 391. doi: 10.1038/msb.2010.52

40. Stoffel W, Jenke B, Block B, Zumbansen M, Koebke J (2005) Neutral sphingomyelinase 2 (*smpd3*) in the control of postnatal growth and development. Proc Natl Acad Sci U S A 102: 4554–4559. doi: 10.1073/pnas.0406380102

41. Stoffel W, Jenke B, Holz B, Binczek E, Gunter RH, et al. (2007) Neutral sphingomyelinase (*SMPD3*) deficiency causes a novel form of chondrodysplasia and dwarfism that is rescued by Col2A1-driven *smpd3* transgene expression. Am J Pathol 171: 153–161. doi: 10.2353/ajpath.2007.061285

42. Kim WJ, Okimoto RA, Purton LE, Goodwin M, Haserlat SM, et al. (2008) Mutations in the neutral sphingomyelinase gene *SMPD3* implicate the ceramide pathway in human leukemias. Blood 111: 4716–4722. doi: 10.1182/blood-2007-10-113068

43. Graw J (2009) Genetics of crystallins: cataract and beyond. Exp Eye Res 88: 173–189. doi: 10.1016/j.exer.2008.10.011

44. Cohen RT, Raby BA, Van Steen K, Fuhlbrigge AL, Celedon JC, et al. (2010) In utero smoke exposure and impaired response to inhaled corticosteroids in children with asthma. J Allergy Clin Immunol 126: 491–497. doi: 10.1016/j.jaci.2010.06.016

45. Plume JM, Beach SR, Brody GH, Philibert RA (2012) A cross-platform genome-wide comparison of the relationship of promoter DNA methylation to gene expression. Front Genet 3: 12. doi: 10.3389/fgene.2012.00012

CHAPTER 5

EVALUATION OF SYSTEMATIC ASSESSMENT OF ASTHMA-LIKE SYMPTOMS AND TOBACCO SMOKE EXPOSURE IN EARLY CHILDHOOD BY WELL-CHILD PROFESSIONALS: A RANDOMIZED TRIAL

ESTHER HAFKAMP-DE GROEN, RALF J. P. VAN DER VALK, ASHNA D. MOHANGOO, JOHANNES C. VAN DER WOUDEN, LIESBETH DUIJTS, VINCENT W. JADDOE, ALBERT HOFMAN, HARRY J. DE KONING, JOHAN C. DE JONGSTE, AND HEIN RAAT

5.1 INTRODUCTION

Asthma is a highly prevalent chronic condition associated with considerable morbidity, reduced health-related quality of life (HRQOL) and significant costs for public health [1], [2]. Interventions aimed at preventing childhood asthma are being developed and evaluated [3]–[9]. While the majority of asthma management education for parents occurs in the clinical setting, increasingly, multifaceted environmental interventions to decrease asthma-like symptoms are delivered by community health workers [7]. Previous studies identified positive outcomes associated with community health worker-delivered interventions, including decreased asthma-like symptoms [7].

In the Netherlands, growth, development and health of all children (0–19 years) is monitored in a nationwide program with regular visits at set ages by well-child care physicians and nurses [10]. The nationwide program is offered free of charge by the government and participation is voluntary (attendance rate ca. 90%) [11]. The well-child care setting creates an opportunity for tailored prevention and promotion of healthy child development. During well-child visits, among other topics that are relevant at the developmental stage of the child, the well-child professionals (medical doctors and nurses) should pay attention to the presence of asthma-like symptoms. However, until now, no systematic assessment of the presence of asthma-like symptoms in early childhood by well-child professionals has been applied at well-child centres in the Netherlands. In the Netherlands, the nationwide well-child care program advises to interview parents regarding environmental tobacco smoke (ETS) exposure to preschool children [11]. However, information leaflets with regard to ETS exposure are not yet given routinely to parents of children aged 1 to 4 years who are exposed to ETS.

This study aimed to evaluate the effectiveness of systematic assessment of asthma-like symptoms and ETS exposure between age 1 and 4 years by well-child professionals. We hypothesised that systematic assessment of asthma-like symptoms and ETS exposure to parents of preschool children (and subsequent counselling such as providing information leaflets or arranging a referral when needed) reduces the prevalence of physician-diagnosed asthma ever and wheezing frequency, and improves fractional exhaled nitric oxide (FeNO, a biomarker of airway inflammation), airway resistance (Rint) and HRQOL measurements at age 6 years. In addition to the study protocol [12], we evaluated whether this approach resulted in a reduction of ETS exposure at home ("ETS exposure at home ever" measured at child age 6 years).

5.2 METHODS

5.2.1 ETHICS STATEMENT

This study is embedded in the Generation R Study, a prospective population-based cohort [13], in collaboration with the regional well-child care organisation Centre for Youth and Family in Rotterdam. The Generation R Study was conducted in accordance with the guidelines proposed in the Declaration of Helsinki, and was approved by the Medical Ethical Committee of the Erasmus Medical Centre. All parents who participated in the Generation R Study provided written informed consent for the use of data regarding their child for research aimed at identifying factors influencing the health of young children. In this study, to evaluate the brief assessment form regarding asthma-like symptoms and ETS exposure applied by well-child professionals, we used data that were collected in the Generation R Study. We are prepared to make the data available upon request.

5.2.2 STUDY DESIGN

Details of our study design were published previously (see File S1) [12]. This study started in June 2005 and follow-up at age 6 years was completed in January 2012. In total, 7775 children (born between April 2002 and January 2006) entered the study (Fig. 1). Sixteen well-child centres that participated in the data collection of the Generation R Study were randomized into 8 well-child centres that applied the brief assessment form regarding asthma-like symptoms and ETS exposure at each regularly scheduled visit to the well-child centre between age 1 and 4 years, and 8 centres that applied usual care. First, the well-child centres were ranked (by researcher ADM) based on the socioeconomic status of their neighbourhood. Well-child centres in each subsequent couple in this list were randomly assigned to the intervention group (n = 8) or the control group (n = 8). Parents were not aware of the research condition they were allocated to. The protocol for this trial and supporting CONSORT checklist are available as supporting information; see Checklist S1 and Protocol S1.

5.2.3 INTERVENTION AND USUAL CARE

When parent and child attended the well-child centre allocated to the intervention group, the professionals used a brief assessment form regarding asthma-like symptoms and ETS exposure during the regular visits at age 14, 24, 36 and 45 months. Details of this form were published previously [12]. In summary, with regard to asthma-like symptoms the brief form included items on wheezing, and shortness of breath or dyspnea. Furthermore, the form included an item that assessed whether the child had been exposed to ETS during the past year (no, yes-sometimes, yes-on a regular basis, yes-often or daily, unknown).

When parents reported that their child had at least 3 episodes of any asthma-like symptoms during the past 12 months and at least 1 episode of asthma-like symptoms in the past 4 weeks, the well-child professionals could provide them with a leaflet with information about asthma. If the child had been free of asthma-like symptoms during the past 4 weeks, the well-child professionals could advise a visit to the general practitioner should the child's asthma-like symptoms return. When parents reported that their child had at least 3 episodes of asthma-like symptoms during the past 12 months, of which at least 1 in the past 4 weeks, and the child had not yet been treated by the general practitioner or paediatrician in the past 4 weeks, the well-child professionals could refer to the asthma nurse and/ or general practitioner. If the child had already been treated by the general practitioner or paediatrician in the past 4 weeks, the well-child professionals could refer to the asthma nurse.

If the child had been exposed to ETS (sometimes, on a regular basis, often or daily), the well-child professional could discuss health risks of ETS exposure to preschool children (health risks), and discuss whether parents could be motivated and prepared to stop ETS exposure to their child (house rules) and provide them with an information leaflet about preventing their child from exposure to ETS. The well-child professionals from the intervention centres were informed during a two-hour session about the intervention.

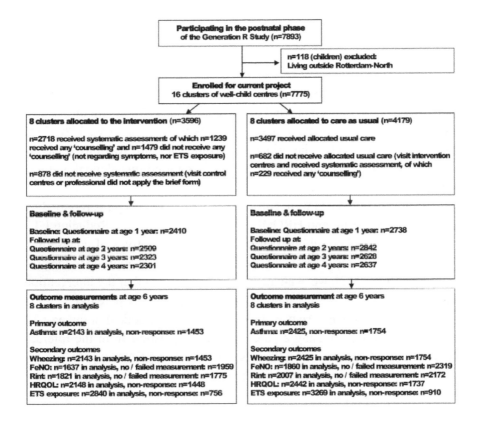

FIGURE 1: Flow of participants through the study. FeNO = fractional exhaled nitric oxide, HRQOL = health related quality of life, Rint = airway resistance, ETS = environmental tobacco smoke.

The control centres applied current routine practice, addressing the presence of general health symptoms during the regular well-child visits and ETS exposure (at least at age 18 months) [11]. However, no specific, systematic assessment of the presence of asthma-like symptoms and ETS exposure by the use of a brief form was performed by the well-child professionals in the control group.

5.2.4 PRIMARY AND SECONDARY OUTCOMES

Data from parents were collected in the Generation R Study by postal questionnaires at enrolment, and at the first, 2nd, 3rd, 4th and 6th year of life. Response rates for these questionnaires were 71%, 76%, 72%, 73% and 68%, respectively. The primary outcome measure was physician-diagnosed asthma ever, obtained by a parent-reported questionnaire at age 6 years.

Secondary outcomes were current wheezing frequency (as reported by parents), FeNO, Rint and HRQOL as reported by parents. Reducing ETS exposure to preschool children was one of the objectives of counselling following systematic assessment of ETS. Therefore, in addition to the proposed outcomes [12], we evaluated at age 6 years whether the intervention had reduced ETS exposure at home ever (as reported by parents).

Wheezing frequency (never, 1–3 episodes, ≥4 episodes) in the past 12 months was assessed using a parent-reported question from the International Study of Asthma and Allergies in Childhood (ISAAC) [14].

FeNO was measured according to American Thoracic Society guidelines [15] at age 6 years at the research centre (NIOX chemiluminescence analyser, Aerocrine AB, Solna, Sweden). Statistical analyses were additionally adjusted for technique to take into account computer-calculated and researcher-observed FeNO values. FeNO was normalized by ^{e}log transformation.

At age 6 years, Rint (Micro Rint, MicroMedical, Rochester, Kent, UK) was measured at the research centre during tidal breathing, with occlusion of the airway at tidal peak expiratory flow. Median values for at least 5 acceptable Rint measurements were calculated and used to calculate Z-scores, additionally adjusted for median variation of the study period [16], [17].

The CHQ-PF28 in the parent-reported questionnaire was used to measure HRQOL of the child at age 6 years [18]. Based on 28 items, the CHQ-PF28 measures the HRQOL of children and their families across 13 scales [19], [20]. The following eight multi-item scales measure the child's HRQOL: Physical functioning, Role functioning: emotional, Role functioning: physical, Bodily pain, General behaviour, Mental health, Self-esteem, General health perceptions. These multi-item scales were summarised into a Physical summary measure and a Psychosocial summary measure. Furthermore we used the Change in health item. The impact of the child's health on the caregiver's and family's HRQOL was measured across the remaining four multi-item scales: Parental impact: emotional, Parental impact: time, Family cohesion and Family activities. All scale measures were transformed to scores ranging from 0 to 100. Lower scores correspond to lower HRQOL. Summary measures were standardised with a mean of 50 and standard deviation of 10 to reflect general US population norms for children [19], [20].

The outcome "ETS exposure at home ever (yes, no)" at age 6 years was defined and based on parent-reported questionnaires at age 2, 3 and 6 years, using the question: "Do people smoke occasionally at home? (yes, no)." "ETS exposure at home ever" at age 6 years was scored "yes" if there was ETS exposure at home at age 2 or 3 or 6 years.

5.2.5 COVARIATES

We used information collected in the Generation R Study on maternal characteristics (educational level, net household income, ethnicity, single motherhood and history of asthma or atopy) for the intervention and control group. Information about the highest attained maternal educational level (low, moderate, high), maternal ethnicity (Dutch, other western, non-western) and single motherhood (yes, no) and maternal history of asthma or atopy (yes, no) were obtained at enrolment by questionnaires. Maternal educational level and maternal ethnicity were defined according to the classification of Statistics Netherlands [21], [22]. Data on household income (<€1600/month, ≥€1600/month) was obtained at the child's age of 3 years, using the 2005 monthly general labour income as the cut-off point [23]. Information on child's gender (boy, girl), gestational age at birth (weeks) and birth weight (grams), were obtained from medical records. We used information collected in the Generation R Study on child's characteristics that were established using parent-reported questionnaires which included: ETS exposure at home (yes, no) (reported during pregnancy) [24]; breastfeeding ever at age 0–6 months (yes, no); keeping pets (yes, no) at the 1st year of life; respiratory tract infections (yes, no) and wheezing (yes, no) at the 1st year of life.

5.2.6 STATISTICAL ANALYSES

Baseline data for the intervention and control group were described using descriptive statistics, which were tested for differences using multinomial regression adjusted for randomisation stratum (cluster). All participants were analysed according to the "intention-to-treat" principle.

The prevalence of ETS exposure at home before (fetal life to age 6 months), during (at age 14–45 months) and after (at age 6 years) the study period was described. P values for differences in the prevalence of 'ETS exposure at home' between intervention and control group were calculated by means of the Chi-square test. Although not according to the study protocol, several children participating in the control group also visited the intervention centres and assessment of asthma-like symptoms and ETS exposure by a brief form was applied to a part of the parents of these children. Contamination of intervention and control condition may possibly also have occurred by moving to another neighbourhood in the city and visiting another well-child centre. Because this contamination may have reduced the differences in results between intervention and control group, we amended the study protocol [12] and in addition to the intention-to-treat analyses we performed a per-protocol analysis. In the per-protocol analysis we included children who were allocated to the intervention group and also received the allocated intervention (n = 2718). In the control group only children were included when they were allocated to the control group and received usual care (n = 3497) (see Fig. 1). Outcomes at age 6 years were predicted with a model using two predictors: research condition (intervention or usual care) and baseline value of the outcome variable [25], [26].

To prevent bias associated with attrition, missing data at baseline and missing outcomes were multiple imputed (10 imputed datasets) on the basis of the correlation between each variable with missing values and other parental and child characteristics [27] to reduce bias and improve efficiency [28]. Regression analyses were performed in the original data and after the multiple imputation procedure. Since we found similar effect estimates (with and without multiple imputation) the final results in our paper are presented as effect estimates with its 95% Confidence Intervals (95%CI) with adjustment for randomisation stratum, derived from the original (unimputed) data. Multilevel regression analyses were applied to allow for dependency between the individual measurements within the 16 randomised well-child centres. (the GENLINMIXED procedure in SPSS and PROC GLIMMIX procedure in SAS) [29], [30]. We considered two levels: the cluster level (well-child center) and the individual(child) level. In the final model, we used the default covariance structure in the multilevel regression analysis in SPSS. The difference between intervention and control group on the categorical outcomes "physician-diagnosed asthma ever (yes/no)" and "ETS exposure at home (yes/no)" were studied using the "binomial" distribution and link = logit. The difference between intervention and control group on the categorical outcome "Wheezing frequency (never, 1–3 times/year, >3 times/year)" was studied using the "multinomial" distribution and link = logit. The differences between intervention and control group on the health-related quality of life scales were studied using the 'poisson' distribution and link = log. The differences between intervention and control group on the outcomes FeNO and Rint were studied using the 'normal' distribution and link = identity. FeNO was normalized by ^{e}log transformation.

Potential effect modification of socio-demographic characteristics and baseline values of the outcomes on the association between the research condition (intervention or care as usual group) and the outcomes was explored. First, we fit a multinomial regression model with randomisation stratum and baseline values of the outcome. Second, we added socio-demographic characteristics (child's gender and maternal ethnic background and educational level) and baseline values of the outcomes as an interaction separately [12], [31], [32]. The interaction terms were evaluated at p<0.10 level [33].

Random treatment allocation ensures that intervention status will not be confounded with either measured or unmeasured baseline characteristics [34]. Therefore, the effect of the intervention on outcomes was estimated by comparing outcomes between the intervention and control group, only adjusted for randomisation stratum and baseline prevalence of the outcomes.

It should be considered that given multiple comparisons, there is an 1-in-20 chance of a false association for each comparison (Type I error at p = 0.05) [35]. Bonferroni correction was applied to correct for multiple testing (P = 0.05/number of comparisons) [35].

In addition, a process evaluation of the intervention was performed. The study is reported according to the CONSORT standards for reporting RCTs [30], [36]. Analyses were performed using the Statistical Package for Social Sciences (SPSS) version 20.0 for Windows (SPSS Inc, Chicago, IL, USA) and SAS 9.2 (SAS Institute, Cary, NC, USA).

5.3 RESULTS

5.3.1 RECRUITMENT

There were 8 intervention and 8 control well-child centres, involving 3596 and 4179 children (and their parents) visiting these well-child centres, respectively. The different rates of participation of the children in the different elements of the study are shown in the flow diagram (Fig. 1).

Table 1 summarizes the baseline characteristics of the study population, stratified by intervention and control group. At baseline, no differences were found between the characteristics of the intervention and control group, after adjustment for randomisation stratum (p>0.05).

TABLE 1: Baseline characteristics by allocation group (n = 7775).

	Missing	Total N = 775 16 clusters	Intervention n - 3596 (46.3%) 8 clusters	Care as usual n = 4719 (53.7%) 8 clusters	P value*
Maternal characteristics					
Educational level	732 (9.4)				
Low		1610 (22.9)	717 (21.8)	893 (23.8)	0.96
Middle		2081 (29.5)	954 (29.0)	1127 (30.0)	
High		3352 (47.6)	1617 (49.2)	1735 (46.2)	
Net household income	2101 (27.0)				
<1600 €/month		1536 (27.1)	608 (23.6)	928 (29.9)	0.56
≥1600 €/month		4138 (72.9)	1966 (76.4)	2172 (70.1)	
Ethnicity	736 (9.5)				
Dutch		3817 (54.2)	1884 (57.4)	1933 (51.5)	0.48
Other western		1186 (16.8)	498 (15.2)	688 (18.3)	
Non-western		2036 (28.2)	900 (27.4)	1136 (30.2)	
Single motherhood (yes)	892 (11.5)	865 (12.6)	408 (12.7)	457 (12.4)	0.93
Smoking during pregnancy (yes)	1717 (22.1)	1510 (24.9)	679 (24.5)	831 (25.3)	0.40
History of asthma or atopy (yes)	1608 (20.7)	2402 (38.9)	1140 (39.1)	1262 (38.8)	0.80
Child's characteristics					
Gender (male)	0 (0)	3920 (50.4)	1796 (49.9)	2124 (50.8)	0.44
Gestational age at birth	0 (0)				
<37 weeks		472 (6.1)	208 (5.8)	264 (6.3)	0.35
≥37 weeks		7303 (93.9)	3388 (94.2)	3915 (93.7)	
Birth weight (grams)	0 (0)				
<2500 grams		438 (5.6)	189 (5.3)	249 (6.0)	0.24
≥2500 grams		7337 (94.4)	3407 (94.7)	3930 (94.0)	
Breastfeeding ever (yes)	1830 (23.5)	6143 (91.9)	2819 (90.6)	3324 (92.9)	0.22
Keeping pets (yes)	2198 (28.3)	1850 (33.2)	872 (33.2)	978 (33.1)	0.66
ETS exposure at home (yes)	3542 (45.6)	662 (15.6)	313 (15.4)	349 (15.8)	0.99
Respiratory tract infections (yes)	2632 (33.9)	3230 (62.8)	1512 (62.8)	1718 (62.8)	0.84
Wheezing (yes)	2860 (36.8)	1482 (30.2)	691 (30.0)	791 (30.3)	0.83

TABLE 1: *Cont.*

Values are absolute numbers (percentages) for categorical variables. *Tested for differences in characteristics in intervention and control group using multinomial regression adjusted for randomization stratum. Characteristic established using postal questionnaires during pregnancy included: smoking during pregnancy (yes, no), maternal atopy (yes, no), maternal ethnicity (Dutch, non-Western, other-Western) and maternal education level. The Dutch Standard Classification of Education was used to categorize women's self-reported highest education qualification [21]: low (less than 4 years of high school), middle (college), and high (Bachelor's degree, Master's degree). Data on net household income were available at the 2nd year of life. Birth weight (grams) and gestational age at birth (weeks) were obtained from medical records. Postnatal factors were established using questionnaires and included: breastfeeding ever at age 0 to 6 months (yes, no); keeping pets (yes, no) at the 1st year of life; ETS exposure at home (yes, no) measured at age 0 to 6 months; respiratory tract infections (yes, no) and wheezing (yes, no) at the 1st year of life.

5.3.1 ASTHMA (RELATED) OUTCOMES

At age 6 years, multilevel regression analysis indicated no differences in asthma, wheezing frequency, FeNO or Rint measurements between the intervention and control group (p>0.05) (Table 2 and 3).

5.3.2 HRQOL

The response rate regarding the CHQ-PF28 scales at age 6 years was different for each scale and varied between 57–59% (n = 4410–4590). Baseline measurements were available for 8 out of 13 CHQ-PF28 scales. At age 6 years, no differences in HRQOL were found between the intervention and control group, after adjustment for baseline HRQOL and randomisation stratum (p>0.05) (Table 2 and 3).

5.3.3 ETS EXPOSURE: BASELINE TO FOLLOW-UP

Figure 2 shows the prevalence of ETS exposure at home before (fetal life to age 6 months), during (at age 14–45 months) and after (at age 6 years) the study period (according to the intention-to-treat analysis). During fetal life and at age 6 months, the prevalence of ETS exposure at home was around

16% in both the intervention and control group (p>0.05). At age 2 years, ETS exposure at home to children participating in the intervention group remained similar, but increased to 19% in the control group. At age 2, 3 and 6 years, the prevalence of ETS exposure at home was higher in children participating in the control group (age 2 years: p = 0.02, age 3 years: p = 0.004, age 6 years: p>0.05). No differences in ETS exposure at home at age 2 and 3 years were found between intervention and control group after adjustment for baseline ETS exposure at home (reported during fetal life) using multinomial regression in an intention-to-treat analysis, (adjusted Odds Ratio [aOR] = 0.90, 95% Confidence Interval [CI]:0.74–1.08 at age 2 years and aOR = 0.81, 95% CI:0.66–1.01 at age 3 years). However, in the per-protocol analysis (n = 1560), multinomial regression analysis indicated a decreased risk on ETS exposure at home in the intervention group at age 2 and 3 years (aOR = 0.78, 95% CI:0.63–0.96 at age 2 years and aOR = 0.73, 95% CI:0.57–0.93 at age 3 years).

5.3.4 ETS EXPOSURE: OUTCOME

At age 6 years, no differences between intervention and control group were found on the outcome "ETS exposure at home ever" using multilevel regression in an intention-to-treat analysis including adjustment for baseline ETS exposure at home (reported during fetal life) (aOR = 0.82, 95% CI:0.66–1.03) (Table 2). However, in an explorative per-protocol analysis, children who received the intervention at the intervention well-child centres had a decreased risk on "ETS exposure at home ever" compared to children who visited the control well-child centres and who did not receive the intervention (aOR = 0.71, 95% CI:0.59–0.87) (Table 3).

5.3.5 INTERACTIONS

No interaction effects on the outcomes were found of the research condition (intervention or control group) with socio-demographic characteristics or baseline values of the outcomes (p>0.10) (data not shown). We found no effect of the frequency of the intervention on outcomes.

TABLE 2: Intention-to-treat analyses: Prevalence and effect estimates of primary and secondary outcomes at age 6 years follow-up by allocation group.

	Intervention n = 3596	Care as usual n = 4179	Adjusted effect estimates [95% CI]*
Primary outcome at age 6 years			
Physician-diagnosed asthma ever[a]	86/2143 (4.0)	101/2425 (4.2)	1.01 (0.76–1.35)
Secondary outcomes at age 6 years			
Wheezing frequency[a]			
Never	1958/2143 (91.4)	2215/2425 (91.3)	Reference
1–3 times/year	143/2143 (6.7)	157/2425 (6.5)	1.02 (0.79, 1.31)
>3 times/year	42/2143 (2.0)	53/2425 (2.2)	0.99 (0.71, 1.37)
Health-related quality of life (CHQ-PF28 scales)[b]			
Physical functioning	97.30 ± 11.16	97.22 ± 11.17	0.00 (–0.01, 0.01)
Role functioning: emotional/behavior	97.40 ± 10.78	97.59 ± 10.28	0.00 (–0.01, 0.00)
Role functioning: physical[d]	97.34 ± 11.41	97.34 ± 11.64	0.00 (–0.01, 0.01)
Bodily pain	86.46 ± 16.71	85.96 ± 17.47	0.00 (–0.01, 0.02)
General behavior[d]	70.72 ± 15.20	71.44 ± 14.68	0.00 (–0.02, 0.03)
Mental health[d]	81.65 ± 14.53	81.90 ± 14.43	0.00 (–0.02, 0.02)
Self esteem[d]	83.81 ± 15.31	83.35 ± 15.28	0.01 (–0.01, 0.03)
General health perceptions	87.19 ± 15.82	86.78 ± 15.74	0.00 (–0.02, 0.02)
Parental impact: emotional	88.76 ± 14.89	89.06 ± 14.52	–0.01 (–0.02, 0.01)
Parental impact: time	95.83 ± 11.89	95.36 ± 13.12	0.00 (–0.01, 0.01)
Family activities	90.81 ± 16.34	90.50 ± 16.23	0.00 (–0.01, 0.01)
Family cohesion	76.31 ± 18.99	76.25 ± 17.94	0.00 (–0.03, 0.02)
Change in health[d]	56.15 ± 15.46	56.84 ± 16.28	–0.01 (–0.06, 0.04)
Physical summary score[d]	57.36 ± 6.22	57.19 ± 6.29	0.17 (–0.58, 0.93)
Psychosocial summary score[d]	53.03 ± 6.79	53.08 ± 6.66	–0.08 (–0.53, 0.37)
FeNO[c,d]	7.20 (0.10–101.00)	7.30 (0.10–119.00)	–0.01 (–0.06, 0.03)
Rint[c,d]	0.93 (0.13–2.43)	0.93 (0.19–2.32)	0.09 (–0.17, 0.35)
ETS exposure at home[a]	567/2840 (20.0)	745/3269 (22.8)	0.82 (0.66, 1.03)

[a] Data are numerator/denominator (%). [b] Mean ± standard deviation. [c] median (range). [d] No baseline measurement available. Numbers of children does not equal the sum of the demoninators in each subgroup because only those with baseline and follow-up data are included. Measurements on FeNO and Rint were available for respectively 3497 (45%) and 3828 (49%) of the participating children. FeNO = Fractional exhaled Nitric Oxide, Rint = airway resistance, ETS = Environmental Tobacco Smoke. *Adjusted for randomization stratum, and baseline prevalence of outcomes. Care as usual is the reference group.

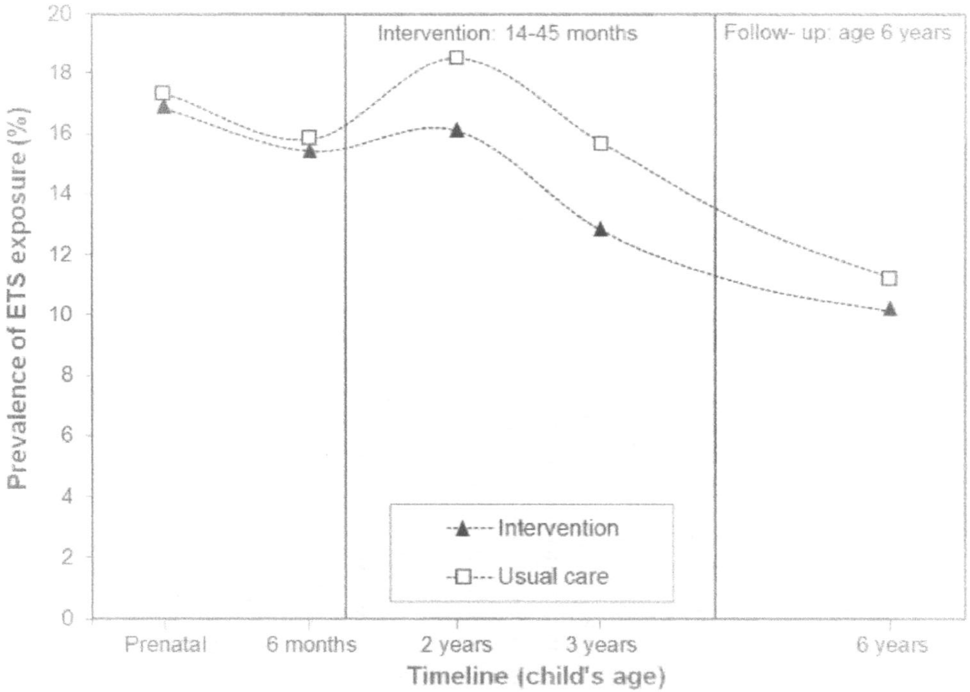

FIGURE 2: Prevalence of ETS exposure at home of intervention and control (usual care) group by child's age (Intention-to-treat analysis). ETS exposure at home was defined based on the question 'Do people smoke occasionally at home?'. Values are percentages and were tested for differences in characteristics in intervention and control group using logistic regression analyses. Population for analysis (N) and P values: Prenatal (N = 5598): p>0.05, 6 months (N = 4233): p>0.05, age 2 years (N = 5290): p = 0.02, age 3 years (N = 4894): p = 0.004, age 6 years (N = 4604): p>0.05.

5.3.6 PROCESS EVALUATION OF THE INTERVENTION

In total, professionals at well-child centres completed 6826 forms to assess asthma-like symptoms and ETS exposure for 2718 children (75.6% of the 3596 children) participating in the intervention group; and 1566 forms were completed for 682 children (16.3% of the 4179 children) participating in the control group (see discussion). In half of the children participating in the intervention group, the brief assessment form was applied at age 14 months (online repository Table S1). In total, the brief assessment form was never applied to 25% of the children participating in the intervention group. To 12% of the children participating in the intervention group, the brief assessment form was applied at each regularly scheduled visit up to year 4 (online repository Table S2).

Of the children in the intervention group who had ≥3 episodes of asthma-like symptoms in the past year, based on the data of the assessment forms, 53% (162/308) was already treated by general practitioner or paediatrician. Of the children with ≥3 episodes of asthma-like symptoms in the past year and asthma-like symptoms during the past month, 86% (119/139) was already treated by general practitioner or paediatrician.

Using the assessment forms, well-child professionals in the intervention group reported a decreasing prevalence of ETS exposure to children participating in the intervention group with increasing child's age: 19% (276/1447) at the age of 14 months, 16% (266/1627) at age 24 months, 17% (301/1767) at age 36 months and 13% (225/1760) at age 45 months. At age 14 months, 89% (245/276) of the children with ETS exposure received the information leaflet regarding the prevention of ETS exposure. However, after the first year, the information leaflet regarding prevention of ETS exposure was less often provided to the parents of children who were exposed to ETS: 61% (163/266) at age 24 months, 64% (192/301) at age 36 months and 53% (119/225) at age 45 months.

5.4 DISCUSSION

Systematic assessment of asthma-like symptoms and ETS exposure by professionals at well-child centres, followed by counselling (when indicated - including referral to asthma nurse/general practitioner and providing parents with information leaflets on avoiding ETS exposure) did not lead to a lower prevalence of physician-diagnosed asthma ever, reduction in parent-reported wheezing symptoms and did not improve FeNO, Rint or parent-reported HRQOL at age 6 years. A decreased risk on ETS exposure at home in the intervention group was found at age 2 and 3 years, but at age 6 years no difference between intervention and control group was found. Process evaluation results showed that most children with wheezing were already treated by their general practitioner or by a paediatrician. Further, half of the parents of children with ETS exposure participating in the intervention group did not receive the information leaflets on ETS exposure at the intervention centres at age 45 months.

This is a community health worker-delivered intervention study using physician-diagnosed asthma ever, wheezing frequency, FeNO, Rint, HRQOL and (in addition) ETS exposure at home ever at age 6 years as outcomes. In contrast to the positive outcomes associated with community health worker-delivered interventions (including decreased asthma-like symptoms) reported by Postma et al [7], our study did not show a lower prevalence of asthma or wheezing after follow-up until age 6 years. Maybe more intensive counselling or interventions based on social cognitive theory, are required to achieve an effect on the asthma related outcomes. By using FeNO and Rint as outcomes we could evaluate the effect of the intervention on airway inflammation and lung function at age 6 years [37], [38], but no effect could be demonstrated. No differences in parent-reported HRQOL were found between intervention and control group, which possibly can be explained by the fact that the intervention did not reduce wheezing.

In addition to the review by Priest et al [39], showing that intensive and repeated counselling interventions seem to be promising to reduce ETS exposure, we found a transient effect of brief counselling aimed to avoid ETS exposure in children at preschool age. To increase efficiency of well-child visits, low intensive and brief assessments and health promotion interventions are preferred. However, process evaluation results showed that half of the parents of children with ETS exposure did not receive the information leaflet regarding prevention of ETS exposure at age 45 months. Apparently, for unknown reason, once prevention of ETS exposure was applied at the first year of life, professionals at well-child care did not tend to repeat the intervention later on while repeated feedback seems to be most effective to reduce the proportion of parents quitting smoking [40], [41].

The strengths of this study include the integration in current practice with a brief assessment form regarding asthma-like symptoms and ETS exposure, the large number of parents participating, the longitudinal design (with follow-up until child age 6 years) and large number of FeNO and Rint measurements. Limitations include shortcomings in the application of the brief assessment forms and counselling. Possible reasons are falling attendance of parents to the well-child centre; lack of time or priority is given to other health questions during the well-child visit or professionals who are not familiar with the intervention, that is still not routine practice. In this study, the professionals were provided with a two-hour specific training on how to apply and use the brief assessment form regarding asthma-like symptoms and ETS exposure. This level of instruction may not be optimal as we did not organize refreshment sessions nor provided feedback on performance or assessed its effect [42].

The study faced some difficulties. In contrast to what was described in our study protocol [12], data on inhaled steroids prescribed by a physician was not available at age 6 years. Asthma at age 6 years was defined as physician-diagnosed asthma ever, obtained by a parent reported questionnaire. In the future, at child's age 10 years, data on inhaled steroids will be available and we recommend repeating the analyses at age 10 years.

In addition to the proposed outcomes, we evaluated whether the intervention had reduced ETS exposure at home. Children participating in the control group also visited the intervention well-child centres and systematic assessment and (when indicated) counselling of asthma-like symptoms and ETS exposure was applied to the parents of these children. Contamination of intervention and control condition may possibly have occurred by moving to another neighbourhood in the city and visiting another well-child centre. Because this contamination may have reduced the differences in results between intervention and control group, we amended the study protocol and in addition to the intention-to-treat analysis we performed a per-protocol analysis.

The following limitations would be a possible explanation for the negative study results: the study included a relatively low-intensity counselling intervention. However, the systematic assessment of the presence of asthma-like symptoms in early childhood by well-child professionals was prioritized and was considered feasible and essential in the Dutch youth healthcare system [43]. Another explanation for the negative study results is that there may have been a lack of intervention by the well-child care professional, and also by the parents/children (to only 12% of the children participating in the intervention group, the brief assessment form was applied at each regularly scheduled visit up to year 4 (Table S2)). Finally, since we used parent reports regarding the presence of asthma symptoms, HRQOL and ETS exposure at home, we may have lost precision.

We consider selection bias unlikely because a multiple imputed analysis including all eligible children did not change the results. Information bias should be considered for different measurements. Although the validity of assessing ETS exposure by questionnaires in epidemiological studies has been shown, misclassification may occur due to underreporting [44]. However, the use of biomarkers of tobacco smoke exposure in urine, saliva or blood, or nicotine in indoor air seems not superior to self-report [44]–[47]. We have to take into account the impact of parental symptom perception and, possibly, misclassification in their reports on asthma diagnosis and symptoms. Parental reports of wheezing are widely accepted in epidemiological studies and reliably reflects the incidence of wheezing in preschool children [14]. However, some misclassification cannot be excluded [48].

The decreased risk on "ETS exposure at home ever" in the intervention group remained statistically significant even after correction for multiple testing.

This study raises questions about whether it is feasible to prevent the development of asthma by using systematic assessment and counselling of asthma-like symptoms and ETS exposure by using brief forms at well-child centres. We recommend further studies to evaluate whether professionals at well-child centres can contribute to optimal asthma management in other ways, and efforts are needed to optimize the protocols that can be implemented in this setting.

We also recommend further studies to improve the current intervention to optimise asthma management at well-child care. Based on previous results, it is recommended that professionals at well-child centres encourage breastfeeding and advise parents of children at high-risk of developing asthma to avoid ETS and indoor allergens exposure to their children to reduce the prevalence of asthma [3], [49]. To optimise asthma management and realise uniformity of practice at well-child care, future opportunities are the development of an assessment to estimate the risk of developing asthma at school age [50]. Further, we stress the importance to ban smoking in public places and residential settings to reduce children's exposure to tobacco smoke.

Our study was embedded within the Dutch system of preventive health care provided by well-child centres in Rotterdam, the Netherlands. This may have consequences for the generalisability of our results in other areas and countries and therefore evaluation of our study in other, varied populations is recommended.

5.5 CONCLUSION

A systematic assessment of asthma-like symptoms and ETS exposure by using brief assessment forms at well-child centres was not effective in reducing the prevalence of physician-diagnosed asthma ever and wheezing, and did not improve FeNO, Rint or HRQOL at age 6 years. Our results hold promise for interviewing parents and using information leaflets at well-child centres to reduce ETS exposure at home in preschool children.

REFERENCES

1. Lai CK, Beasley R, Crane J, Foliaki S, Shah J, et al. (2009) Global variation in the prevalence and severity of asthma symptoms: phase three of the International Study of Asthma and Allergies in Childhood (ISAAC). Thorax 64: 476–483. doi: 10.1136/thx.2008.106609
2. Masoli M, Fabian D, Holt S, Beasley R, Global Initiative for Asthma P (2004) The global burden of asthma: executive summary of the GINA Dissemination Committee report. Allergy 59: 469–478. doi: 10.1111/j.1398-9995.2004.00526.x

3. Chan-Yeung M, Ferguson A, Watson W, Dimich-Ward H, Rousseau R, et al. (2005) The Canadian Childhood Asthma Primary Prevention Study: outcomes at 7 years of age. J Allergy Clin Immunol 116: 49–55. doi: 10.1016/j.jaci.2005.03.029

4. Howden-Chapman P, Pierse N, Nicholls S, Gillespie-Bennett J, Viggers H, et al. (2008) Effects of improved home heating on asthma in community dwelling children: randomised controlled trial. BMJ 337: a1411. doi: 10.1136/bmj.a1411

5. Lanphear BP, Aligne CA, Auinger P, Weitzman M, Byrd RS (2001) Residential exposures associated with asthma in US children. Pediatrics 107: 505–511. doi: 10.1542/peds.107.3.505

6. Pilotto LS, Nitschke M, Smith BJ, Pisaniello D, Ruffin RE, et al. (2004) Randomized controlled trial of unflued gas heater replacement on respiratory health of asthmatic schoolchildren. Int J Epidemiol 33: 208–214. doi: 10.1093/ije/dyh018

7. Postma J, Karr C, Kieckhefer G (2009) Community health workers and environmental interventions for children with asthma: a systematic review. J Asthma 46: 564–576. doi: 10.1080/02770900902912638

8. Schonberger HJ, Dompeling E, Knottnerus JA, Maas T, Muris JW, et al. (2005) The PREVASC study: the clinical effect of a multifaceted educational intervention to prevent childhood asthma. Eur Respir J 25: 660–670. doi: 10.1183/09031936.05.00067704

9. Arshad SH, Bateman B, Matthews SM (2003) Primary prevention of asthma and atopy during childhood by allergen avoidance in infancy: a randomised controlled study. Thorax 58: 489–493. doi: 10.1136/thorax.58.6.489

10. Burgmeijer RJF, Van Geenhuizen YM, Filedt Kok-Weimar T, De Jager AM (1997) Op weg naar volwassenheid, Evaluatie Jeugdgezondheidszorg 1996. Leiden/Maarssen: TNO/KPMG.

11. Ministry of Health Welfare and Sport (2003) Basistakenpakket Jeugdgezondheidszorg 0–19 jaar. Den Haag: Ministry of HWS.

12. Hafkamp-de Groen E, Mohangoo AD, de Jongste JC, van der Wouden JC, Moll HA, et al. (2010) Early detection and counselling intervention of asthma symptoms in preschool children: study design of a cluster randomised controlled trial. BMC Public Health 10: 555. doi: 10.1186/1471-2458-10-555

13. Jaddoe VW, van Duijn CM, Franco OH, van der Heijden AJ, van IIzendoorn MH, et al. (2012) The Generation R Study: design and cohort update 2012. Eur J Epidemiol 27: 739–756. doi: 10.1007/s10654-012-9735-1

14. Jenkins MA, Clarke JR, Carlin JB, Robertson CF, Hopper JL, et al. (1996) Validation of questionnaire and bronchial hyperresponsiveness against respiratory physician assessment in the diagnosis of asthma. Int J Epidemiol 25: 609–616. doi: 10.1093/ije/25.3.609

15. American Thoracic Society, European Respiratory Society (2005) ATS/ERS recommendations for standardized procedures for the online and offline measurement of exhaled lower respiratory nitric oxide and nasal nitric oxide, 2005. Am J Respir Crit Care Med 171: 912–930. doi: 10.1164/rccm.200406-710st

16. Merkus PJ, Stocks J, Beydon N, Lombardi E, Jones M, et al. (2010) Reference ranges for interrupter resistance technique: the Asthma UK Initiative. Eur Respir J 36: 157–163. doi: 10.1183/09031936.00125009

17. Van der Valk RJ, Kiefte-de Jong JC, Sonnenschein-van der Voort AM, Duijts L, Hafkamp-de Groen E, et al. (2013) Neonatal folate, homocysteine, vitamin B12 levels and methylenetetrahydrofolate reductase variants in childhood asthma and eczema. Allergy 68: 788–795. doi: 10.1111/all.12146

18. Raat H, Botterweck AM, Landgraf JM, Hoogeveen WC, Essink-Bot ML (2005) Reliability and validity of the short form of the child health questionnaire for parents (CHQ-PF28) in large random school based and general population samples. J Epidemiol Community Health 59: 75–82. doi: 10.1136/jech.2003.012914

19. Landgraf J, Abetz J, Ware JE (1999) Child Health Questionnaire (CHQ): A User's Manual. Boston, MA: HealthAct.

20. HealthActCHQ (2008) Child health Questionnaire Scoring and Interpretation Manual. HealthActCHQ Inc., Cambridge MA, USA.

21. Statistics Netherlands (2004) The Dutch Standard Classification of Education. Voorburg/Heerlen: Statistics Netherlands.

22. Swertz O, Duimelaar P, Thijssen J (2004) Migrants in the Netherlands 2004. Voorburg/Heerlen: Statistics Netherlands.

23. CPB Netherlands Bureau for Economic Policy Analysis (2005) Beschrijving koopkrachtberekening, CPB Memorandum 133. Available: http://www.cpb.nl/en/publication/beschrijving-koopkrachtberekening. Accessed 2013 March 4.

24. Duijts L, Jaddoe VW, van der Valk RJ, Henderson JA, Hofman A, et al. (2012) Fetal exposure to maternal and paternal smoking and the risks of wheezing in preschool children: the Generation R Study. Chest 141: 876–885. doi: 10.1378/chest.11-0112

25. Twisk J, Proper K (2004) Evaluation of the results of a randomized controlled trial: how to define changes between baseline and follow-up. J Clin Epidemiol 57: 223–228. doi: 10.1016/j.jclinepi.2003.07.009

26. Vickers AJ, Altman DG (2001) Statistics notes: Analysing controlled trials with baseline and follow up measurements. BMJ 323: 1123–1124. doi: 10.1136/bmj.323.7321.1123

27. Sterne JA, White IR, Carlin JB, Spratt M, Royston P, et al. (2009) Multiple imputation for missing data in epidemiological and clinical research: potential and pitfalls. BMJ 338: b2393. doi: 10.1136/bmj.b2393

28. Spratt M, Carpenter J, Sterne JA, Carlin JB, Heron J, et al. (2010) Strategies for multiple imputation in longitudinal studies. Am J Epidemiol 172: 478–487. doi: 10.1093/aje/kwq137

29. Twisk J (2006) Applied Multilevel Analysis: A Practical Guide. Cambridge: Cambridge University Press.

30. Campbell MK, Elbourne DR, Altman DG (2004) group C CONSORT statement: extension to cluster randomised trials. BMJ 328: 702–708. doi: 10.1136/bmj.328.7441.702

31. Assmann SF, Pocock SJ, Enos LE, Kasten LE (2000) Subgroup analysis and other (mis)uses of baseline data in clinical trials. Lancet 355: 1064–1069. doi: 10.1016/s0140-6736(00)02039-0

32. Pocock SJ, Assmann SE, Enos LE, Kasten LE (2002) Subgroup analysis, covariate adjustment and baseline comparisons in clinical trial reporting: current practice and problems. Stat Med 21: 2917–2930. doi: 10.1002/sim.1296

33. Sun X, Briel M, Walter SD, Guyatt GH (2010) Is a subgroup effect believable? Updating criteria to evaluate the credibility of subgroup analyses. BMJ 340: c117. doi: 10.1136/bmj.c117

34. Greenland S, Pearl J, Robins JM (1999) Causal diagrams for epidemiologic research. Epidemiology 10: 37–48. doi: 10.1097/00001648-199901000-00008

35. Wit E, McClure J (2004) Statistics for Microarrays: design, analysis and inference. Chichester: Wiley & Sons, Ltd.

36. Altman DG, Schulz KF, Moher D, Egger M, Davidoff F, et al. (2001) The revised CONSORT statement for reporting randomized trials: explanation and elaboration. Ann Intern Med 134: 663–694. doi: 10.7326/0003-4819-134-8-200104170-00012

37. Beydon N, Pin I, Matran R, Chaussain M, Boule M, et al. (2003) Pulmonary function tests in preschool children with asthma. Am J Respir Crit Care Med 168: 640–644. doi: 10.1164/rccm.200303-449oc

38. Taylor DR, Pijnenburg MW, Smith AD, De Jongste JC (2006) Exhaled nitric oxide measurements: clinical application and interpretation. Thorax 61: 817–827. doi: 10.1136/thx.2005.056093

39. Priest N, Roseby R, Waters E, Polnay A, Campbell R, et al.. (2008) Family and carer smoking control programmes for reducing children's exposure to environmental tobacco smoke. Cochrane Database Syst Rev: CD001746.

40. Wilson SR, Farber HJ, Knowles SB, Lavori PW (2011) A randomized trial of parental behavioral counseling and cotinine feedback for lowering environmental tobacco smoke exposure in children with asthma: results of the LET'S Manage Asthma trial. Chest 139: 581–590. doi: 10.1378/chest.10-0772

41. Wilson SR, Yamada EG, Sudhakar R, Roberto L, Mannino D, et al. (2001) A controlled trial of an environmental tobacco smoke reduction intervention in low-income children with asthma. Chest 120: 1709–1722. doi: 10.1378/chest.120.5.1709

42. Emmons KM, Rollnick S (2001) Motivational interviewing in health care settings. Opportunities and limitations. Am J Prev Med 20: 68–74. doi: 10.1016/s0749-3797(00)00254-3

43. Dijkstra NS, Van Wijngaarden JCM, De Vries J, Lim-Feijen JF, Raat H, et al.. (2001) Programmeringstudie effectonderzoek Jeugdgezondheidszorg. Utrecht: GGD Nederland.

44. Patrick DL, Cheadle A, Thompson DC, Diehr P, Koepsell T, et al. (1994) The validity of self-reported smoking: a review and meta-analysis. Am J Public Health 84: 1086–1093. doi: 10.2105/ajph.84.7.1086

45. Brunekreef B, Leaderer BP, van Strien R, Oldenwening M, Smit HA, et al. (2000) Using nicotine measurements and parental reports to assess indoor air: the PIAMA birth cohort study. Prevention and Incidence of Asthma and Mite Allergy. Epidemiology 11: 350–352. doi: 10.1097/00001648-200005000-00023

46. Margolis PA, Keyes LL, Greenberg RA, Bauman KE, LaVange LM (1997) Urinary cotinine and parent history (questionnaire) as indicators of passive smoking and predictors of lower respiratory illness in infants. Pediatr Pulmonol 23: 417–423. doi: 10.1002/(sici)1099-0496(199706)23:6<417::aid-ppul4>3.0.co;2-f

47. Wang X, Tager IB, Van Vunakis H, Speizer FE, Hanrahan JP (1997) Maternal smoking during pregnancy, urine cotinine concentrations, and birth outcomes. A prospective cohort study. Int J Epidemiol 26: 978–988. doi: 10.1093/ije/26.5.978

48. Cane RS, Ranganathan SC, McKenzie SA (2000) What do parents of wheezy children understand by "wheeze"? Arch Dis Child 82: 327–332. doi: 10.1136/adc.82.4.327
49. Becker A, Watson W, Ferguson A, Dimich-Ward H, Chan-Yeung M (2004) The Canadian asthma primary prevention study: outcomes at 2 years of age. J Allergy Clin Immunol 113: 650–656. doi: 10.1016/j.jaci.2004.01.754
50. Hafkamp-de Groen E, Lingsma HF, Caudri D, Wijga A, Jaddoe VW, et al. (2012) Predicting asthma in preschool children with asthma symptoms: study rationale and design. BMC Pulm Med 12: 65. doi: 10.1186/1471-2466-12-65

There is one table and several supplemental files that are not available in this version of the article. To view this additional information, please use the citation on the first page of this chapter.

PART III

AMBIENT AND HOUSEHOLD EXPOSURES

CHAPTER 6

PRENATAL EXPOSURE TO PERSISTENT ORGANOCHLORINES AND CHILDHOOD OBESITY IN THE U.S. COLLABORATIVE PERINATAL PROJECT

LEA A. CUPUL-UICAB, MARK A. KLEBANOFF, JOHN W. BROCK, AND MATTHEW P. LONGNECKER

6.1 INTRODUCTION

Childhood obesity is of public health concern worldwide (Lobstein et al. 2004). In the United States, the prevalence of obesity among children 6–11 years of age increased from 4% in 1971–1974 to 20% in 2007–2008 (Orsi et al. 2011). Obesity during childhood is predictive of obesity in adulthood (Orsi et al. 2011). Similar to obese adults, children who are obese are at increased risk of developing adverse health conditions such as insulin resistance, type 2 diabetes, hypertension, dyslipidemia, and asthma (Lobstein et al. 2004). Excess energy intake and low levels of physical activity are well documented risk factors for childhood obesity (Lobstein et al. 2004). The role of an adverse fetal environment (e.g., maternal diabetes,

Used with permission from Environmental Health Perspectives. Cupul-Uicab LA, Klebanoff MA, Brock JW, and Longnecker MP. Prenatal Exposure to Persistent Organochlorines and Childhood Obesity in the U.S. Collaborative Perinatal Project. Environmental Health Perspectives, 121 (2013), http://dx.doi.org/10.1289/ehp.1205901.

malnutrition, smoking) in the programming of childhood obesity is supported by previous studies (Huang et al. 2007; Oken and Gillman 2003). Emerging literature from animal models also suggests that prenatal exposure to endocrine-disrupting chemicals (e.g., tributyltin) might predispose the exposed subjects to greater accumulation of body fat later in life (Grun and Blumberg 2009).

Organochlorine compounds are manufactured chemicals that were widely used as pesticides [e.g., DDT (dichlorodiphenyltrichloroethane), aldrin, dieldrin, HCB (hexachlorobenzene)] and in industrial processes [PCBs (polychlorinated biphenyls)] between the 1950s and 1980s; they are ubiquitous in the environment, tend to bioaccumulate, and have a high affinity for fatty tissues [Centers for Disease Control and Prevention (CDC) 2009]. Organochlorines at high levels are known to be toxic for wildlife and humans. The use of these chemicals is presently banned or restricted; but because of the chemicals' persistence in the environment, adverse health outcomes related to background levels of exposure are still a concern for the general population (World Health Organization 2010). In humans, prenatal exposure to some persistent organochlorines such as p,p′-DDE [1,1-dichloro-2,2-bis(p-chlorophenyl)ethylene, the main breakdown product of DDT], PCBs, and HCB has been associated with higher body weight or body mass index (BMI; kilograms per meter squared) across a wide range of ages; however, overall the results from previous studies are equivocal (Tang-Peronard et al. 2011).

The main purpose of the present study was to assess the association between prenatal exposure to persistent organochlorines [i.e., β-hexachlorocyclohexane (β-HCH), p,p′-DDE, p,p′-DDT, dieldrin, heptachlor epoxide, hexachlorobenzene, trans-nonachlor, oxychlordane, and PCBs] and subsequent childhood obesity in a relatively large population whose exposure levels were measured in maternal serum collected before these chemicals were banned in the United States. Potential interactions reported by earlier studies (e.g., p,p′-DDE and maternal smoking) were explored using child's BMI as the outcome. We also evaluated whether any associations appeared to differ according to size at birth, which might mediate the relationship between prenatal exposures and childhood body weight.

6.2 METHODS

The present study was based on the U.S. Collaborative Perinatal Project (CPP). Detailed information about the CPP is available elsewhere (Niswander and Gordon 1972). Briefly, the CPP was a prospective cohort that enrolled pregnant women from 1959 to 1965 at 12 study centers across the United States. Women provided nonfasting blood samples throughout their pregnancy and at delivery; serum was stored in glass at −20°C with no recorded thaws. The CPP enrolled about 42,000 women who delivered 55,000 babies. These children were systematically assessed for various outcomes until 7 years of age; the follow-up rate at this age was about 75%. Exposure to organochlorines was measured in a subset of the CPP cohort; children were eligible if they were singleton and live-born, and there was a third trimester maternal serum sample of 3 mL available. Of the 44,075 children who met the eligibility criteria, 1,200 were selected at random and 1,623 were selected according to sex-specific birth defects or performance on various neurodevelopmental tests (Longnecker et al. 2001). The sampling was done independently, and the 71 children selected for more than one subgroup were included in the analysis only once. The study complied with all applicable requirements of the U.S. regulations to conduct research in humans, women gave oral informed consent prior to the (CPP) study, and the Institutional Review Board of the National Institute of Environmental Health Sciences deemed that the present analysis was exempt.

Organochlorine exposure. Maternal serum samples were analyzed for β-HCH, p,p′-DDE, p,p′-DDT, dieldrin, heptachlor epoxide, HCB, trans-nonachlor, oxychlordane, and 11 PCB congeners (28, 52, 74, 105, 118, 138, 153, 170, 180, 194, and 203) at the CDC from 1997 to 1999. Quantification of these organochlorines was done using electron capture detection after solid-phase extraction, cleanup, and dual-column gas chromatography (Brock et al. 1996). Measured values reported by the laboratory that were below the limit of detection (LOD) were used in the analyses; no imputation of values below the LOD was done (Longnecker et al. 2002). The average proportion of chemicals recovered by extraction ranged between 45 and 80; for most the recovery was > 60% (i.e., recovery was < 60% for PCB congeners 170, 194, and 203) (Brock et al. 1996). The results are shown without recovery adjustment (Longnecker et al. 2002). For the present analysis, the concentrations of the 11 PCB congeners were summed to calculate total PCBs. Serum triglycerides and total cholesterol (milligrams per deciliter) were measured with standard enzymatic methods (Longnecker et al. 2002).

Organochlorines were measured in third- rather than first-trimester maternal serum, which is potentially the critical period for developmental exposure. However, previous data from the CPP have shown that serum levels of persistent organochlorines (i.e., p,p'-DDE and PCBs) early in pregnancy correlated well with those found at the end of pregnancy (Longnecker et al. 1999).

Outcomes. At the 7-year examination, children had their height and weight measured by trained personnel as part of the CPP study. Childhood overweight and obesity were defined using age- and sex-specific cut points for BMI as recommended by the International Obesity Task Force (IOTF), which allows prevalence comparisons across countries (Cole et al. 2000). These cut points correspond to a BMI approximately above the 90th centile for overweight and approximately above the 97th centile for obesity (specific cut points defined by 6-month age bands for girls and boys, and ranged from 17.53 to 18.44 for overweight and from 20.08 to 21.60 for obesity) (Cole et al. 2000). The IOTF cut points are considered equivalent to the well-established cut points for BMI to define overweight (25 kg/m^2) and obesity (30 kg/m^2) in adults (Cole et al. 2000). The primary outcome for the present analysis was childhood obesity. Because the number of children classified as obese was small, overweight including obesity (hereafter referred to as overweight unless otherwise noted) and BMI as a continuous variable were also modeled as secondary outcomes.

Covariates. Socioeconomic index was defined based on education level and occupation of the head of the household and the family income at enrollment (Longnecker et al. 2002). Maternal smoking during pregnancy was ascertained at enrollment. Pregnancy weight gain (kilograms) was calculated from self-reported prepregnancy weight and weight measured at the end of pregnancy. Maternal prepregnancy BMI was calculated from self-reported prepregnancy weight and measured height. Information about breastfeeding was recorded in the nursery wards and refers to whether the child was breastfed during the first week of life.

Organochlorine levels were missing for 3% of the 2,823 children selected, mostly because the measured value did not meet the quality control standards for acceptance (Brock et al. 1996). The present analysis was restricted to children whose organochlorine exposure was measured and those with available height, weight, and exact age for the 7-year examination (n = 2,094). Children with implausible values for height or weight (n = 13) and those missing data on maternal prepregnancy BMI (n = 144), smoking (n = 11), pregnancy weight gain (n = 8), or birth order (n = 2) were excluded, leaving 1,916 children with data for analysis. The number of children included in analyses of each organochlorine ranged between 1,807 and 1,915 because of varying numbers of missing values for each of them.

Statistical analysis. The association of prenatal exposure to organochlorines with childhood obesity and childhood overweight was modeled with logistic regression. Separate models were fitted for each of the nine chemicals and each outcome. The shape of the relationships between the outcomes and exposures was examined using restricted cubic splines (of the exposure) with three knots (Harrell 2001), a method for modeling nonlinear relationships. In the spline models, odds ratios (ORs) for overweight and obesity were estimated using the lowest level of each organochlorine as the reference (Orsini and Greenland 2011). ORs for overweight and obesity were also estimated per interquartile range (IQR) increase in organochlorine levels (micrograms per liter). All models were weighted by the inverse of the sampling probabilities to account for the sampling design used to select the subset of CPP with organochlorine measures (Zhou et al. 2007). All analyses were conducted using STATA (release 12.1; StataCorp, College Station, TX, USA).

We used directed acyclic graphs (DAGs) to choose potential confounders for model adjustment (Greenland et al. 1999; Textor et al. 2011). The set of variables selected for adjustment were maternal race, education, socioeconomic index, prepregnancy BMI, smoking during pregnancy, and child's birth order. This list of variables plus study center (12 strata), child's sex, serum levels of triglycerides and total cholesterol were the covariates selected a priori for model adjustment. Another set of potential confounders was also assessed using the change in estimate method (i.e., change in OR ≥ 10%), starting with all variables in the models with deletion of one by one in a stepwise manner (Greenland 1989). None of the additionally tested variables (i.e., maternal age, pregnancy weight gain, weeks of gestation at enrollment, and child's age) were selected with this strategy.

Potential interactions of the exposure with breastfeeding (child was breastfed at the nursery ≥ 1 days: no, yes), maternal smoking (no, yes), child's sex (male, female), child born small for gestational age [SGA, defined as a birth weight for gestational age below the 10th percentile (no, yes)], and preterm birth [child born before 37 weeks of gestation (no, yes)] were examined by introducing cross-product terms for these variables [e.g., organochlorine (micrograms per liter) × breastfeeding (no, yes)] into the regression models. All interactions were tested in the context of linear regression with BMI as the outcome because of limited statistical power when modeling childhood overweight or obesity. Interactions with a p-value for the cross-term product ≤ 0.20 were further assessed with stratified analyses.

The following set of additional models was run in sensitivity analyses only. We estimated adjusted ORs for overweight and obesity using categories of exposure; the cut points to define categories were selected based on quantiles of distribution specific to each organochlorine, which aimed to keep observations below the LOD as the lowest category; for organochlorines with > 22% values below the LOD (i.e., trans-nonachlor and oxychlordane) we used tertiles and for the others we used quintiles. For each analysis, a 1-df (degrees of freedom) trend test was obtained, based on assigning to each subject an exposure value equal to the median organochlorine level among subjects in their respective organochlorine category. Overweight and obesity were also modeled using organochlorine levels expressed per lipid basis (nanograms per gram lipids) as the exposure; in that case, the regression models did not include triglycerides or total cholesterol as covariates. Child's BMI as an outcome was also assessed in relation to organochlorine exposure by fitting linear regression models that were additionally adjusted for child's exact age when the anthropometric measurements were taken. In addition, the analysis was restricted to the children selected at random (i.e., from the subset of the CPP cohort with organochlorine measurements), to those with organochlorine levels above the LOD, to non-breastfed children, and finally to those without intrauterine growth restriction who were also born at term. We also fitted models with additional adjustment for breastfeeding, SGA, and preterm birth. We performed multiple imputation by chained equations (van Buuren et al. 1999) for covariates with missing data, and the analyses were repeated using the imputed data (see Supplemental Material, Multiple Imputation).

TABLE 1: Characteristics of the mothers and children according to child's BMI[a] at 7 years of age in the CPP, 1959–1965.

Characteristics	Normal BMI		Overweight		Obese		p-Value[c]
	n	Percent[b] or median (IQR)	n	Percent[b] or median (IQR)	n	Percent[b] or median (IQR)	
All	1,683	87.8	165	8.6	68	3.5	
Mother							
Age at recruitment (years)	1,683	23.0 (8.0)	165	23.0 (10.0)	68	23.0 (9.0)	0.27
Race							< 0.01
White	725	85.3	90	10.6	35	4.1	
African American	904	90.3	70	7.0	27	2.7	
Other	54	83.1	5	7.7	6	9.2	
Education							0.06
< High school	972	87.6	89	8.0	48	4.3	
≥ High school	711	88.1	76	9.4	20	2.5	
Socioeconomic index							0.03
≤ 3.0	468	90.5	33	6.4	16	3.1	
3.1–6.0	776	86.3	82	9.1	41	4.6	
> 6.0	439	87.8	50	10.0	11	2.2	
Smoking status (cigarettes/day)							0.66
Nonsmoker	932	88.2	93	8.8	32	3.0	
Smoke ≤ 10	469	88.0	43	8.1	21	3.9	
Smoke > 10	282	86.5	29	8.9	15	4.6	
Prepregnancy BMI (kg/m²)							< 0.01
< 25.0	1,335	89.8	110	7.4	42	2.8	
25.0–29.9	243	80.7	40	13.3	18	6.0	
≥ 30.0	105	82.0	15	11.7	8	6.3	
Pregnancy weight gain (kg)	1,683	10.4 (5.4)	165	10.8 (5.9)	68	11.7 (5.2)	0.01
Children							
Sex							0.01

TABLE 1: *Cont.*

Characteristics	Normal BMI		Overweight		Obese		p-Value[c]
	n	Percent[b] or median (IQR)	n	Percent[b] or median (IQR)	n	Percent[b] or median (IQR)	
Male	1,042	88.4	107	9.1	30	2.5	
Female	641	87.0	58	7.9	38	5.2	
Live birth order							0.09
First	502	85.4	57	9.7	29	4.9	
Second	361	87.2	39	9.4	14	3.4	
≥ Third	820	89.7	69	7.5	25	2.7	
Breastfeeding ≥ 1 days							0.17
No	1,300	88.1	119	8.1	56	3.8	
Yes	258	86.3	35	11.7	6	2.0	
Missing	125	88.0	11	7.7	6	4.3	
Birth length (cm)	1,664	50.0 (4.0)	165	50.0 (4.0)	67	50.0 (3.0)	0.27
Birth weight (g)	1,682	3,203 (681)	165	3,317 (709)	68	3,161 (751)	< 0.01
Weight (kg)[d]	1,683	22.7 (3.7)	165	28.5 (5.0)	68	35.3 (6.7)	< 0.01
Height (cm) [d]	1,683	121.0 (7.0)	165	123.0 (9.0)	68	125.0 (7.5)	< 0.01
BMI (kg/m²) [d]	1,683	15.5 (1.7)	165	18.7 (1.0)	68	22.3 (3.1)	< 0.01

[a]*Categories defined by age- and sex-specific cut-off points for BMI from the IOTF.* [b]*Some row percentages do not add up 100% due to rounding.* [c]*p-Values testing differences in percentages (Pearson's chi-square or Fisher's exact test) or medians (Kruskal–Wallis) across the three groups defined by child's BMI.* [d]*At the 7-year examination.*

6.3 RESULTS

The mean (± SD) age of the children in the study was 7.1 ± 0.2 years). The prevalence of childhood overweight (excluding obesity) was 8.6% and of obesity was 3.5%. Higher prevalence of overweight and obesity was observed among children whose mothers were overweight or obese before pregnancy (Table 1). The prevalence of obesity was slightly higher among children of white women than African-American women.

As expected, those organochlorines with higher median concentrations tended to have a greater proportion of measurements ≥ LOD (Table 2); the Spearman correlation coefficient between median concentration and proportion ≥ LOD was 0.99 (calculation based on 19 organochlorines, including specific PCB congeners). The between-assay coefficient of variation

(CV) was < 25% for most of the chemicals, except for HCH (CV, 59%). Compared with adults ≥ 20 years of age from the U.S. National Health and Nutrition Examination Survey (NHANES 2003–2004), women from the CPP had higher levels of exposure to all organochlorines (Figure 1) (CDC 2009). NHANES data showed nondetectable levels of dieldrin, heptachlor epoxide, and p,p′-DDT in 50% of the U.S. adult population; median levels of p,p′-DDE were approximately 17 times higher in the CPP.

TABLE 2: Third-trimester serum levels of organochlorines among women from the CPP, 1959–1965.

Organochlorines	n	CV (%)	LOD (µg/L)	< LOD (%)	Selected percentiles (µg/L)			
					25th	50th	75th	95th
β-HCH	1,899	18.6	0.23	0.2	1.01	1.42	2.12	5.25
p,p'-DDE	1,833	19.2	0.61	0.0	16.93	24.59	36.35	69.63
p,p'-DDT	1,903	22.1	0.66	0.1	6.46	9.33	14.16	26.57
Dieldrin	1,807	19.9	0.23	1.1	0.60	0.81	1.09	1.81
Heptachlor epoxide	1,859	18.6	0.21	19.7	0.24	0.40	0.74	1.53
HCB	1,881	59.3	0.08	21.3	0.12	0.24	0.35	0.82
trans-Nonachlor	1,911	19.7	0.28	28.4	0.26	0.40	0.58	0.91
Oxychlordane	1,809	24.6	0.20	29.8	0.16	0.33	0.53	0.95
Total PCBsa	1,915	18.7	0.20–0.33	5.2–98.4	1.93	2.74	3.92	6.74

CV, coefficient of variation. [a]Congener-specific LOD (µg/L): 74 and 203, 0.20; 105 and 194, 0.21; 170, 0.22; 180, 0.23; 118 and 138, 0.25; 28 and 52, 0.27; 153, 0.33. The proportion below LOD was, for 138, 5.2%; 153, 8.7%; 118, 10.2%; 74, 36.8%; 180, 49.5%; and for the other congeners, ≥ 76.8%.

After adjusting for potential confounders, increasing levels of HCB, heptachlor epoxide, β-HCH, p,p′-DDE, total PCBs, trans-nonachlor, and oxychlordane were not associated with childhood obesity in the present study (Figure 2A). p,p′-DDT was associated with an increased odds of obesity, but this was not statistically significant. Increasing levels of dieldrin were associated with obesity, but the confidence intervals were particularly wide (Figure 2A). None of the organochlorines were associated with overweight (including obesity); in particular, the ORs for dieldrin were close to the null (Figure 2B). An IQR increase in levels of HCB, heptachlor epoxide, β-HCH, p,p'-DDE, p,p'-DDT, total PCBs, trans-non-achlor, and oxychlordane was not associated with overweight, obesity, or BMI; dieldrin was associated with obesity [adjusted OR = 1.32; 95% CI: 1.01, 1.73] but not with overweight or BMI (Table 3). Compared with the lowest quintile of dieldrin exposure, children from higher quintiles had higher odds of obesity (p-trend = 0.08); however, given the small number of children classified as obese (3.5%), the estimates were imprecise as shown by the wide confidence intervals (Table 4). Similar to the results from the restricted cubic splines, no association between quintiles of diel-drin exposure and overweight or BMI were observed.

TABLE 3: Associations between maternal exposure to persistent organochlorines (per IQR increase) and offspring body size in the CPP, 1959–1965.

Chemicals (µg/L)	IQR	Overweight[a]	Obese	BMI (kg/m²)
		OR[b] (95% CI)	OR[b] (95% CI)	β[b,c] (95% CI)
β-HCH	1.11	1.00 (0.87, 1.16)	1.03 (0.84, 1.24)	0.01 (−0.06, 0.09)
p,p'-DDE	19.42	0.88 (0.72, 1.07)	1.00 (0.75, 1.32)	−0.03 (−0.18, 0.11)
p,p'-DDT	7.70	0.97 (0.79, 1.19)	1.16 (0.88, 1.54)	0.02 (−0.12, 0.16)
Dieldrin	0.49	1.00 (0.83, 1.22)	1.32 (1.01, 1.73)	0.09 (−0.12, 0.29)
Heptachlor epoxide	0.50	1.06 (0.82, 1.36)	0.96 (0.48, 1.93)	0.00 (−0.18, 0.18)
HCB	0.23	1.00 (0.98, 1.03)	1.02 (0.99, 1.05)	0.02 (−0.02, 0.07)
trans-Nonachlor	0.32	1.05 (0.82, 1.34)	1.00 (0.67, 1.51)	0.01 (−0.15, 0.17)
Oxychlordane	0.37	0.92 (0.67, 1.26)	1.00 (0.53, 1.90)	−0.01 (−0.20, 0.17)
Total PCBs	1.99	0.97 (0.80, 1.18)	1.01 (0.71, 1.44)	0.01 (−0.12, 0.14)

[a]*Includes overweight and obese.* [b]*Adjusted for total cholesterol, triglycerides, study center, mother's race, socioeconomic index, education, smoking during pregnancy, prepregnancy BMI, and child's sex and birth order.* [c]*Additionally adjusted for child's exact age at anthropometric measurements.*

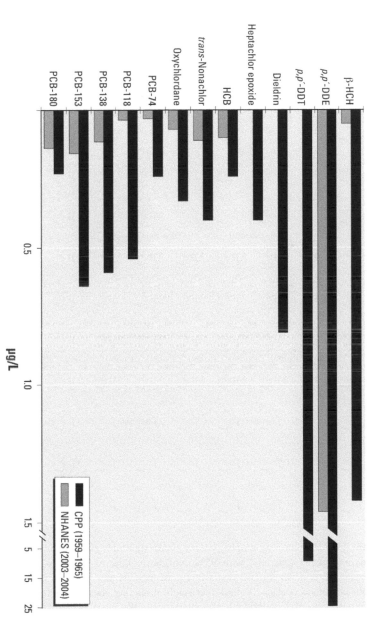

FIGURE 1: Bar graph comparing organochlorine levels between CPP participants and NHANES 2003-04 participants, y-axis = eight different organochlorines and five PCB congeners, x-axis = levels in µg/L. Details in figure legends and Results. Figure 1 – Median levels of organochlorine chemicals measured in adult serum from the NHANES compared with women from the CPP. The NHANES 50th percentile for dieldrin, heptachlor epoxide, and p,p'-DDT was below the LOD (< 0.01 µg/L).

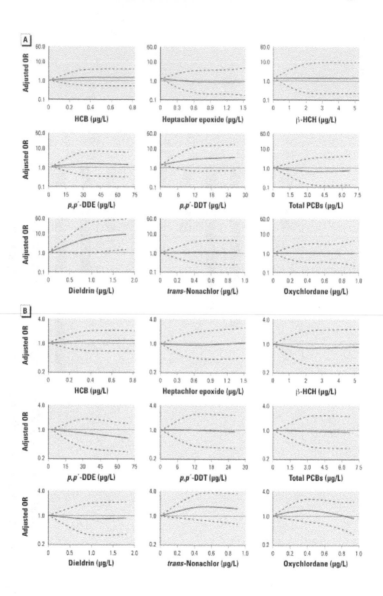

FIGURE 2: Nine line graphs, y-axes = adjusted odds ratios of obesity (Figure 2A) and overweight (Figure 2B) with 95% confidence interval bands, x-axes = levels of each organochlorine in μg/L. Details in figure legends and Results.Figure 2. Adjusted ORs for childhood obesity (A) and overweight including obesity (B) across levels of prenatal exposure to each individual organochlorine. ORs were estimated from restricted cubic splines with three knots, the reference is the lowest level of each organochlorine. The x-axes were truncated at the 95th percentile of the distribution for all organochlorines. Dotted lines represent the 95% CIs.

TABLE 4: Associations between maternal exposure to dieldrin and offspring's body size in the CPP, 1959–1965.

Dieldrin (μg/L)[a]	Overweight[b]		Obese		BMI (kg/m²)
	n	OR (95% CI)[c]	n	OR (95% CI)[c]	β (95% CI)[c,d]
< 0.57	50	1.00	8	1.00	0.00
0.57–0.72	37	0.78 (0.44, 1.39)	8	1.49 (0.46, 4.85)	0.07 (–0.23, 0.38)
0.73–0.91	46	0.84 (0.48, 1.48)	12	1.39 (0.47, 4.15)	0.12 (–0.16, 0.41)
0.92–1.18	41	0.83 (0.46, 1.48)	20	3.62 (1.25, 10.46)	0.22 (–0.15, 0.59)
> 1.18	48	0.79 (0.44, 1.43)	16	2.31 (0.76, 7.09)	0.13 (–0.24, 0.50)
p-Trend[e]		0.61		0.08	0.50

[a]Quintiles. [b]Includes overweight and obese. [c]Adjusted for total cholesterol, triglycerides, study center, mother's race, socioeconomic index, education, smoking during pregnancy, prepregnancy BMI, and child's sex and birth order. [d]Additionally adjusted for child's exact age at anthropometric measurements. [e]Trend test from a model that included quintiles of dieldrin as an ordinal variable, median level of each quintile defined each category.

Potential interactions (p-interaction ≤ 0.20) were observed for maternal smoking and total PCBs; for child's sex and p,p′-DDT, dieldrin, heptachlor epoxide, and oxychlordane; for breastfeeding and β-HCH, p,p′-DDE, p,p′-DDT, dieldrin, and HCB; and for children born SGA with β-HCH and HCB. After stratification, however, there was little evidence of effect modification (see Supplemental Material, Tables S1–S4). Briefly, PCBs (per interquartile increase) were associated with higher child's BMI among smoking mothers and lower child's BMI among nonsmoking mothers (see Supplemental Material, Table S1). Increasing levels of p,p′-DDT, dieldrin, heptachlor epoxide, and oxychlordane tended to be associated with higher BMI among boys but lower BMI among girls (see Supplemental Material, Table S2). We were unable to stratify on breastfeeding (most children were not breastfed) and SGA because of the small numbers. However, estimates among non-breastfed children and among those with a birth weight > 10th percentile for gestational week were comparable to those in Table 3 (see Supplemental Material, Table S3, S4).

The results did not change materially when the models were refitted using categories of exposure (p-values for trend test were all ≥ 0.24) or when organochlorines were expressed per lipid basis (nanograms per gram lipids) (data not shown); and as before, dieldrin was associated with obesity [adjusted OR = 1.42; 95% CI: 1.01, 1.85 (per IQR increase, 64 ng/g lipid)] but not with overweight or BMI. After restricting the analyses to the group of children selected at random (i.e., from the subset of CPP cohort with organochlorine measurements) or to those with organochlorine levels above the LOD, results were consistent with those observed in Table 3 (data not shown). Among children from the random sample, the adjusted OR for obesity per interquartile increase in dieldrin was 1.21 (95% CI: 0.91, 1.61). After excluding children born preterm or SGA, some of the ORs for obesity changed by > 10% (e.g., oxychlordane: OR = 1.31; 95% CI: 0.69, 2.49; see Supplemental Material, Table S5], however, the estimates were imprecise and the association with dieldrin remained. Restricting the analysis to non-breastfeed children (the majority; only 299 children were breastfed) showed results comparable to those from Table 3 (see Supplemental Material, Table S6). The results did not materially change from those in Table 3 after additional adjustment for potentially intermediate variables (i.e., SGA, premature birth, and breastfeeding) (data not shown). The results from the multiple imputation analyses showed that the ORs for obesity were reduced by 5% when dieldrin was modeled as continuous (per IQR increase) and by 30–37% when using quintiles; otherwise, they were comparable to those based on complete data (data not shown).

6.4 DISCUSSION

Overall, prenatal exposure to persistent organochlorines was not associated with obesity, overweight, or BMI among children from the CPP. Increasing levels of dieldrin exposure, however, showed an association with childhood obesity, but it was not associated with the other two outcomes (i.e., overweight or BMI). This finding was consistent among children in the random sample; however, in that sample the adjusted OR for obesity did not reach statistical significance, likely due to the smaller number of obese children among this subset. Nevertheless, a chance finding cannot be ruled out given the number of exposures (n = 9) and outcomes (n = 3) assessed in the present study (for the main analyses that includes the assessment of interactions, ~ 81 models were fitted).

We are not aware of any previous study of the potential role of prenatal exposure to dieldrin in the development of obesity in humans. However, acute aldrin exposure in pregnant mice resulted in decreased body weight of the offspring [Agency for Toxic Substances and Disease Registry (ATSDR) 2002]. [Aldrin is converted to dieldrin in vivo (ATSDR 2002).] Recent data show low levels of serum dieldrin (75th percentile, 0.06 μg/L) among the U.S. population compared with the CPP participants (75th percentile, 1.09 μg/L) (CDC 2009).

A summary of prior studies assessing prenatal exposure to β-HCH, p,p′-DDE, p,p′-DDT, HCB, and PCBs in relation to body weight or BMI is presented in Supplemental Material, Table S7. Similar to previous studies conducted among subjects who experienced relatively high exposure, prenatal exposure to p,p′-DDE was not associated with BMI (Cupul-Uicab et al. 2010; Gladen et al. 2004) (see Supplemental Material, Table S7). Earlier studies that reported an association between DDE exposure and higher BMI during infancy and childhood (Mendez et al. 2011; Valvi et al. 2012; Verhulst et al. 2009) had median levels of p,p′-DDE that were lower than or close to the lowest level of exposure found in the CPP participants (lowest value of p,p′-DDE in the CPP, 0.34 μg/g lipid). In addition, these non-null findings tended to be limited to certain subgroups (i.e., higher BMI with DDE exposure among children whose mothers smoked ever or among children whose mothers had a normal prepregnancy BMI) (Mendez et al. 2011; Verhulst et al. 2009). In the present study, however, DDE did not show an interaction with maternal smoking or prepregnancy BMI.

Similar to our results, two previous studies that assessed prenatal exposure to p,p′-DDT in relation to BMI also reported null findings among infants and youth (Cupul-Uicab et al. 2010; Gladen et al. 2004) (see Supplemental Material, Table S7).

An association between prenatal exposure to HCB and BMI or obesity was not evident in the present study. The previous study that supported such an association was based on a population with levels of HCB (measured in cord blood) higher than those found in the CPP (Smink et al. 2008) (see Supplemental Material, Table S7).

As observed in the present study, null associations between prenatal PCBs exposure and subsequent weight or BMI (i.e., infants and adults) have been previously reported among populations with PCBs levels that are higher or lower than or similar to those found in the CPP participants (Jackson et al. 2010; Karmaus et al. 2009; Patandin et al. 1998) (see Supplemental Material, Table S7). However, our results differ from those of earlier studies that have linked higher levels of PCBs exposure with decreased weight among children and youth, which included populations with PCBs levels that were similar to or higher than the CPP (Blanck et al. 2002; Jacobson et al. 1990; Lamb et al. 2006).

As noted previously with DDE, prenatal exposure to DDT and PCBs was associated with overweight or high BMI only in studies that included populations with low levels of exposure (see Supplemental Material, Table S7), unlike the CPP. Potential effect modification of DDE and PCBs by sex, reported by earlier studies (Gladen et al. 2000; Hertz-Picciotto et al. 2005; Lamb et al. 2006), was not supported in the CPP (see Supplemental Material, Table S2, S7).

Overall, prior studies evaluating developmental exposure to persistent organochlorines in relation to body size have provided little support for the hypothesis that higher levels of exposure are linked to obesity; yet with the available evidence, an association cannot be ruled out (see Supplemental Material, Table S7). Conflicting findings across studies might be explained by methodological variations and by particular characteristics of the studied populations. However, it is also plausible that prenatal exposure to these chemicals at low levels, but not at higher levels, may promote adiposity (Grun and Blumberg 2009).

The mechanism(s) by which prenatal exposure to persistent organochlorine chemicals might be related to increased body fatness are not clear. Potential pathways might entail alterations of the hormones involved in growth regulation and adipogenesis (i.e., thyroid, steroids, and growth hormone) (Cocchi et al. 2009; Diamanti-Kandarakis et al. 2009; Garten et al. 2012; Grun and Blumberg 2009) or of regulation of behavior by the central nervous system (ATSDR 2002).

Compared with previous studies evaluating prenatal exposure to persistent organochlorines in relation to body weight or BMI, the present analysis included by far the largest number of subjects (see Supplemental

Material, Table S7). Although some of the CPP children included in our study were selected based on sex-specific birth defects or their performance on various neurodevelopmental tests, the findings from the analysis that included all subjects (which accounted for selection design) were maintained when the analyses were restricted to children selected at random. Whereas the selection status of the children may be associated with some of the organochlorines, selection status was not associated with obesity, overweight, or BMI. A serious selection bias due to the inclusion of children with sex-specific birth defects or neurodevelopmental delays was therefore unlikely.

In the present study, levels of all organochlorines were much higher than those reported recently for the U.S. population (Figure 1). CPP participants experienced particularly high exposure to DDT and its main breakdown product, p,p′-DDE. Few people with very low levels of p,p′-DDE were in the CPP, thus comparing results to studies with statistically significant findings (Mendez et al. 2011; Verhulst et al. 2009) is not straightforward.

Compared with children followed to 7 years of age, those lost to follow-up had slightly higher median levels of p,p′-DDE (25.7 μg/L), dieldrin (0.87 μg/L), and trans-nonachlor (0.44 μg/L), but they had slightly lower levels of heptachlor epoxide (0.37 μg/L), HCB (0.21 μg/L), and PCBs (2.55 μg/L). This pattern shows that non-followed children did not consistently have higher levels of exposure; furthermore, given the small differences in exposure between children who were or were not followed, selection bias seems unlikely.

Children from the CPP were prospectively followed, and their height and weight were measured by trained personnel who did not know the child's exposure status; thus, differential misclassification of the outcomes was unlikely. Using obesity as an outcome may be a better surrogate measure of adiposity than BMI because high BMI among relatively thin children may not reflect excess body fatness (Freedman and Sherry 2009). Because there are no unified criteria to classify children as overweight or obese, these two outcomes were defined using the IOTF criteria (age- and sex-specific cut points) that allow comparison of the prevalence across countries (Cole et al. 2000). Among 6- to 8-year-olds, the IOTF cut points to classify overweight children have high sensitivity (> 83%) and specificity (> 91%), but the cut points to classify obese children have low sensitivity (68%) and high specificity (> 98%) compared with percentage of body fat (Zimmermann et al. 2004). Using percentage of body fat as the gold standard, the IOTF criteria to define overweight and obesity in children 6–8 years of age showed lower sensitivity but higher specificity than the CDC criteria (Zimmermann et al. 2004).

Compared with recent data from U.S. children (Orsi et al. 2011), the prevalence of overweight and obesity among children in the CPP was low, which decreased the statistical power of the study, especially to assess interactions when modeling these outcomes. All interactions were tested using BMI as the outcome, as was done in previous studies with smaller samples sizes; however, previously reported interactions (i.e., exposure with maternal smoking and child's sex) were not replicated in the present study. Thus, it is not clear whether the exposure has differential effects on child's BMI as previously suggested. Our data did not support interactions between prenatal exposure to organochlorines and fetal growth restriction (i.e., SGA) or prematurity; however, among children without fetal growth restriction who were born at term, organochlorine exposure was not associated with body size at 7 years, except for dieldrin and obesity, as before (see Supplemental Material, Table S5).

Except for HCB, the between-assay CV was relatively low for all organochlorines measured in the present study. Although for some chemicals the percentage of samples below the LOD was 20–30%, our results were unchanged when the analysis was restricted to the sample with levels above the LOD.

6.5 CONCLUSIONS

In the present study, which included a relatively large sample of children with comparatively high prenatal exposure to persistent organochlorines, no clear associations emerged between exposure and obesity or BMI. The suggestive association between dieldrin and childhood obesity was perhaps a chance finding given the number of analyses we performed. However, the present data do not refute a potential role of prenatal exposure to persistent organochlorines in the development of obesity. Low levels of exposure to persistent organochlorines in the prenatal period could conceivably promote obesity, but overall these data provide little support for an association at higher levels.

REFERENCES

1. ATSDR (Agency for Toxic Substances and Disease Registry). 2002. Toxicological Profile for Aldrin/Dieldrin (Update). Atlanta, GA:Agency for Toxic Substances and Disease Registry.

2. Blanck HM, Marcus M, Rubin C, Tolbert PE, Hertzberg VS, Henderson AK, et al. 2002. Growth in girls exposed in utero and postnatally to polybrominated biphenyls and polychlorinated biphenyls. Epidemiology 13(2):205–210.

3. Brock JW, Burse VW, Ashley DL, Najam AR, Green VE, Korver MP, et al. 1996. An improved analysis for chlorinated pesticides and polychlorinated biphenyls (PCBs) in human and bovine sera using solid-phase extraction. J Anal Toxicol 20(7):528–536.

4. CDC (Centers for Disease Control and Prevention). 2009. Fourth Report on Human Exposure to Environmental Chemicals. Atlanta, GA:Centers for Disease Control and Prevention.

5. Cocchi D, Tulipano G, Colciago A, Sibilia V, Pagani F, Viganò D, et al. 2009. Chronic treatment with polychlorinated biphenyls (PCB) during pregnancy and lactation in the rat: Part 1: Effects on somatic growth, growth hormone-axis activity and bone mass in the offspring. Toxicol Appl Pharmacol 237(2):127–136.

6. Cole TJ, Bellizzi MC, Flegal KM, Dietz WH. 2000. Establishing a standard definition for child overweight and obesity worldwide: international survey. BMJ 320(7244):1240–1243.

7. Cupul-Uicab LA, Hernández-Avila M, Terrazas-Medina EA, Pennell ML, Longnecker MP. 2010. Prenatal exposure to the major DDT metabolite 1,1-dichloro-2,2-bis(p-chlorophenyl)ethylene (DDE) and growth in boys from Mexico. Environ Res 110(6):595–603.

8. Diamanti-Kandarakis E, Bourguignon JP, Giudice LC, Hauser R, Prins GS, Soto AM, et al. 2009. Endocrine-disrupting chemicals: an Endocrine Society scientific statement. Endocr Rev 30(4):293–342.

9. Freedman DS, Sherry B. 2009. The validity of BMI as an indicator of body fatness and risk among children. Pediatrics 124(suppl 1):S23–S34.

10. Garten A, Schuster S, Kiess W. 2012. The insulin-like growth factors in adipogenesis and obesity. Endocrinol Metab Clin North Am 41(2):283–295.

11. Gladen BC, Klebanoff MA, Hediger ML, Katz SH, Barr DB, Davis MD, et al. 2004. Prenatal DDT exposure in relation to anthropometric and pubertal measures in adolescent males. Environ Health Perspect 112:1761–1767; doi:10.1289/ehp.7287.

12. Gladen BC, Ragan NB, Rogan WJ. 2000. Pubertal growth and development and prenatal and lactational exposure to polychlorinated biphenyls and dichlorodiphenyl dichloroethene. J Pediatr 136(4):490–496.

13. Greenland S. 1989. Modeling and variable selection in epidemiologic analysis. Am J Public Health 79(3):340–349.
14. Greenland S, Pearl J, Robins JM. 1999. Causal diagrams for epidemiologic research. Epidemiology 10(1):37–48.
15. Grun F, Blumberg B. 2009. Endocrine disrupters as obesogens. Molecular and cellular endocrinology 304(1–2):19–29.
16. Harrell FEJ. 2001. Regression Modeling Strategies: With Applications to Linear Models, Logistic Regression, and Survival Analysis. New York:Springer.
17. Hertz-Picciotto I, Charles MJ, James RA, Keller JA, Willman E, Teplin S. 2005. In utero polychlorinated biphenyl exposures in relation to fetal and early childhood growth. Epidemiology 16(5):648–656.
18. Huang JS, Lee TA, Lu MC. 2007. Prenatal programming of childhood overweight and obesity. Matern Child Health J 11(5):461–473.
19. Jackson LW, Lynch CD, Kostyniak PJ, McGuinness BM, Louis GM. 2010. Prenatal and postnatal exposure to polychlorinated biphenyls and child size at 24 months of age. Reprod Toxicol 29(1):25–31.
20. Jacobson JL, Jacobson SW, Humphrey HE. 1990. Effects of exposure to PCBs and related compounds on growth and activity in children. Neurotoxicol Teratol 12(4):319–326.
21. Karmaus W, Osuch JR, Eneli I, Mudd LM, Zhang J, Mikucki D, et al. 2009. Maternal levels of dichlorodiphenyl-dichloroethylene (DDE) may increase weight and body mass index in adult female offspring. Occup Environ Med 66(3):143–149.
22. Lamb MR, Taylor S, Liu X, Wolff MS, Borrell L, Matte TD, et al. 2006. Prenatal exposure to polychlorinated biphenyls and postnatal growth: a structural analysis. Environ Health Perspect 114:779–785; doi:10.1289/ehp.8488.
23. Lobstein T, Baur L, Uauy R. 2004. Obesity in children and young people: a crisis in public health. Obes Rev 5(suppl 1):4–104.
24. Longnecker MP, Klebanoff MA, Brock JW, Zhou H, Gray KA, Needham LL, et al. 2002. Maternal serum level of 1,1-dichloro-2,2-bis(p-chlorophenyl)ethylene and risk of cryptorchidism, hypospadias, and polythelia among male offspring. Am J Epidemiol 155(4):313–322.
25. Longnecker MP, Klebanoff MA, Gladen BC, Berendes HW. 1999. Serial levels of serum organochlorines during pregnancy and postpartum. Arch Environ Health 54(2):110–114.
26. Longnecker MP, Klebanoff MA, Zhou H, Brock JW. 2001. Association between maternal serum concentration of the DDT metabolite DDE and preterm and small-for-gestational-age babies at birth. Lancet 358(9276):110–114.
27. Mendez MA, Garcia-Esteban R, Guxens M, Vrijheid M, Kogevinas M, Goñi F, et al. 2011. Prenatal organochlorine compound exposure, rapid weight gain, and overweight in infancy. Environ Health Perspect 119:272–278; doi:10.1289/ehp.1002169.
28. Niswander KR, Gordon M. 1972. The Women and Their Pregnancies. Washington, DC:National Institutes of Health.
29. Oken E, Gillman MW. 2003. Fetal origins of obesity. Obes Res 11(4):496–506.
30. Orsi CM, Hale DE, Lynch JL. 2011. Pediatric obesity epidemiology. Curr Opin Endocrinol Diabetes Obes 18(1):14–22.

31. Orsini N, Greenland S. 2011. A procedure to tabulate and plot results after flexible modeling of a quantitative covariate. Stata J 11(1):1–29.
32. Patandin S, Koopman-Esseboom C, de Ridder MA, Weisglas-Kuperus N, Sauer PJ. 1998. Effects of environmental exposure to polychlorinated biphenyls and dioxins on birth size and growth in Dutch children. Pediatr Res 44(4):538–545.
33. Smink A, Ribas-Fito N, Garcia R, Torrent M, Mendez MA, Grimalt JO, et al. 2008. Exposure to hexachlorobenzene during pregnancy increases the risk of overweight in children aged 6 years. Acta Paediatr 97(10):1465–1469.
34. Tang-Peronard JL, Andersen HR, Jensen TK, Heitmann BL. 2011. Endocrine-disrupting chemicals and obesity development in humans: a review. Obes Rev 12(8):622–636.
35. Textor J, Hardt J, Knüppel S. 2011. DAGitty: a graphical tool for analyzing causal diagrams. Epidemiology 22(5):745; doi:10.1097/EDE.0b013e318225c2be.
36. Valvi D, Mendez MA, Martinez D, Grimalt JO, Torrent M, Sunyer J, et al. 2012. Prenatal concentrations of polychlorinated biphenyls, DDE, and DDT and overweight in children: a prospective birth cohort study. Environ Health Perspect 120:451–457; doi:10.1289/ehp.1103862.
37. van Buuren S, Boshuizen HC, Knook DL. 1999. Multiple imputation of missing blood pressure covariates in survival analysis. Stat Med 18(6):681–694.
38. Verhulst SL, Nelen V, Hond ED, Koppen G, Beunckens C, Vael C, et al. 2009. Intrauterine exposure to environmental pollutants and body mass index during the first 3 years of life. Environ Health Perspect 117:122–126; doi:10.1289/ehp.0800003.
39. World Health Organization. 2010. Persistent Organic Pollutants: Impact on Child Health. Geneva: World Health Organization. Available: http://www.who.int/ceh/publications/persistent_organic_pollutant/en/index.html [accessed 9 August 2013].
40. Zhou H, Chen J, Rissanen TH, Korrick SA, Hu H, Salonen JT, et al. 2007. Outcome-dependent sampling: an efficient sampling and inference procedure for studies with a continuous outcome. Epidemiology 18(4):461–468.
41. Zimmermann MB, Gubeli C, Puntener C, Molinari L. 2004. Detection of overweight and obesity in a national sample of 6–12-y-old Swiss children: accuracy and validity of reference values for body mass index from the US Centers for Disease Control and Prevention and the International Obesity Task Force. Am J Clin Nutr 79(5):838–843.

There are several supplemental files that are not available in this version of the article. To view this additional information, please use the citation on the first page of this chapter.

CHAPTER 7

BIRTH OUTCOMES AND MATERNAL RESIDENTIAL PROXIMITY TO NATURAL GAS DEVELOPMENT IN RURAL COLORADO

LISA M. MCKENZIE, RUIXIN GUO, ROXANA Z. WITTER, DAVID A. SAVITZ, LEE S. NEWMAN, AND JOHN L. ADGATE

7.1 INTRODUCTION

Approximately 3.3% of U.S. live-born children have a major birth defect (Centers for Disease Control and Prevention 2013; Parker et al. 2010); these defects account for 20% of infant deaths as well as 2.3% of premature death and disability (McKenna et al. 2005). Oral clefts, neural tube defects (NTDs), and congenital heart defects (CHD) are the most common classes of birth defects (Parker et al. 2010). These defects are thought to originate in the first trimester as a result of polygenic inherited disease or gene–environment interactions (Brent 2004). Suspected nongenetic risk factors for these birth defects include folate deficiency (Wald and Sneddon 1991), maternal smoking (Honein et al. 2006), alcohol abuse and solvent use (Romitti et al. 2007), and exposure to benzene (Lupo et al. 2010b; Wennborg et al. 2005), toluene (Bowen et al. 2009), polycyclic aromatic hydrocar-

Used with permission from Environmental Health Perspectives. McKenzie LM, Guo R, Witter RZ, Savitz DA, Newman LS, and Adgate JL. Birth Outcomes and Maternal Residential Proximity to Natural Gas Development in Rural Colorado. Environmental Health Perspectives, *122 (2014), http://dx.doi.org/10.1289/ehp.1306722.*

bons (PAHs) (Ren et al. 2011), and petroleum-based solvents, including aromatic hydrocarbons (Chevrier et al. 1996). Associations between air pollution [volatile organic compounds (VOCs), particulate matter (PM), and nitrogen dioxide (NO_2)] and low birth weight and preterm birth have been reported (Ballester et al. 2010; Brauer et al. 2008; Dadvand et al. 2013; Ghosh et al. 2012; Llop et al. 2010). Many of these air pollutants are emitted during development and production of natural gas (referred to herein as NGD), and concerns have been raised that they may increase risk of adverse birth outcomes and other health effects (Colborn et al. 2011; McKenzie et al. 2012). Increased prevalence of low birth weight and small for gestational age and reduced APGAR scores were reported in infants born to mothers living near NGD in Pennylvania (Hill 2013).

Technological advances in directional drilling and hydraulic fracturing have resulted in a global boom of drilling and production of natural gas reserves [U.S. Energy Information Administration (EIA) 2011a, 2011b; Vidas and Hugman 2008]. NGD is an industrial process resulting in potential worker and community exposure to multiple environmental stressors (Esswein et al. 2013; King 2012; Witter et al. 2013). Diesel-powered heavy equipment is used for worksite development as well as transporting large volumes of water, sand, and chemicals to sites and for waste removal (Witter et al. 2013). It is increasingly common for NGD to encroach on populated areas, potentially exposing more people to air and water emissions as well as to noise and community-level changes that may arise from industrialization [Colorado Oil and Gas Conservation Commission (COGCC) 2009]. Studies in Colorado, Texas, Wyoming, and Oklahoma have demonstrated that NGD results in emission of VOCs, NO_2, sulfur dioxide (SO_2), PM, and PAHs from either the well itself or from associated drilling processes or related infrastructure (i.e., drilling muds, hydraulic fracturing fluids, tanks containing waste water and liquid hydrocarbons, diesel engines, compressor stations, dehydrators, and pipelines) (CDPHE 2007; Frazier 2009; Kemball-Cook et al. 2010; Olaguer 2012; Walther 2011; Zielinska et al. 2011). Some of these pollutants, such as toluene, xylenes, and benzene, are suspected teratogens (Lupo et al. 2010b; Shepard 1995) or mutagens (Agency for Toxic Substances and Disease Registry 2007) and are known to cross the placenta (Bukowski 2001), raising the possibility of fetal exposure to these and other pollutants resulting from

NGD. Currently, there are few studies on the effects of air pollution or NGD on birth outcomes.

In this analysis, we explored the association between maternal exposure to NGD and birth outcomes, using a data set with individual-level birth data and geocoded natural gas well locations. We conducted a retrospective cohort study to investigate the association between density and proximity of natural gas wells within a 10-mile radius of maternal residences in rural Colorado and three classes of birth defects, preterm birth, and fetal growth.

7.2 METHODS

Study population. We used information available in the publically accessible Colorado Oil and Gas Information System (COGIS) to build a geocoded data set with latitude, longitude, and year of development (1996–2009) for all gas wells in rural Colorado (COGIS 2011). Live birth data were obtained from the Colorado Vital Birth Statistics (CDPHE, Denver, CO). Geocoded maternal addresses at time of birth were linked to the well locations. Distance of each maternal residence from all existing (not abandoned) natural gas wells within a 10-mile radius was then computed using spherically adjusted straight line distances. We conducted our analysis on the final de-identified database containing maternal and birth outcome data described below and distance to all wells within the 10-mile radius. The Colorado Multiple Institutional Review Board reviewed and approved our study protocol. Informed consent was not required.

We restricted analysis to births occurring from 1996 through 2009 to focus our analysis on growth of unconventional NGD, characterized by use of hydraulic fracturing and/or directional drilling (King 2012), which expanded rapidly in Colorado beginning around 2000 (COGIS 2011). We also restricted our analysis to rural areas and towns with populations of < 50,000 (excluding the Denver metropolitan area, El Paso County, and the cities of Fort Collins, Boulder, Pueblo, Grand Junction, and Greeley) in 57 counties to reduce potential for exposure to other pollution sources, such as traffic, congestion, and industry. The final study area included locations with and without NGD. We conducted a retrospective study on the resulting cohort of 124,842 live births to explore associations between birth outcomes and exposure to NGD operations. We restricted eligibility to singleton births and excluded the small proportion (< 5%) of nonwhite births because there were too few to analyze separately.

Birth outcomes. Identified birth outcomes were a) oral cleft, including cleft lip with and without cleft palate as well as cleft palate [International Classification of Diseases, Ninth Revision, Clinical Modification (ICD-9-CM) code 749.xx] (National Center for Health Statistics 2011); b) NTD, including anencephalus, spina bifida without anecephaly, and encephalocele (ICD-9-CM 740.xx, 741.xx, and 742.0); c) CHD, including transposition of great vessels, tetralogy of Fallot, ventricular septal defect, endocardial cushion defect, pulmonary valve atresia and stenosis, tricuspid valve atresia and stenosis, Ebstein's anomaly, aortic valve stenosis, hypoplastic left heart syndrome, patent ductus arteriosis, coarctation of aorta, and pulmonary artery anomalies (codes 745.xx, 746.xx, 747.xx, excluding 746.9, 747.5); d) preterm birth (< 37 weeks completed gestation); e) term low birth weight (≥ 37 weeks completed gestation and birth weight < 2,500 g); and f) term birth weight as a continuous measure. Births with an oral cleft, NTD, or CHD were excluded from preterm birth and term low birth weight analysis. Preterm births were excluded from term birth weight analysis. Oral cleft, CHD, and NTD cases in the Colorado Responds to Children with Special Needs (CRCSN) birth registry, obtained from hospital records, the Newborn Genetics Screening Program, the Newborn Hearing Screening Program, laboratories, physicians, and genetic, developmental, and other specialty clinics (CRCSN 2011) were matched with Colorado live birth certificates. Cases are reflective of reporting as of 12 July 2012, were not necessarily confirmed by medical record review, and are subject to change as CRCSN ascertains diagnosis up to 3 years of child's age and/or supplements information by medical record review. We analyzed birth defects in three heterogeneous groups to increase statistical power. Data set information was not sufficient to distinguish between multiple and isolated birth anomalies or to identify chromosomal birth anomalies. In an exploratory analysis, we considered seven clinical diagnostic groupings of CHDs: a) conotruncal defects (tetralogy of Fallot and transposition of great vessels); b) endocardial cushion and mitrovalve defects (EMD; endocardial cushion defect and hypoplastic left heart syndrome); c) pulmonary artery and valve defects (PAV; pulmonary valve atresia and stenosis and pulmonary artery anomalies); d) tricuspid valve defects (TVD; tricuspid valve atresia and stenosis and Ebstein's anomaly); e) aortic artery and valve defects (aortic valve stenosis and co-arctation of aorta); f) ventricular septal defects (VSD); and g) patent ductus arteriosis in births > 2,500 g (Gilboa et al. 2005).

Exposure assessment. Distribution of the wells within a 10-mile radius of maternal residence shows 50% and 90% of wells to be within 2.3 and 7.7 miles of maternal residence, respectively. We used an inverse distance weighted (IDW) approach, commonly used to estimate individual air pollutant exposures from multiple fixed locations (Brauer et al. 1998; Ghosh et al. 2012), to estimate maternal exposure. Our IDW well count accounts for the number of wells within the 10-mile radius of the maternal residence, as well as distance of each well from the maternal residence, giving greater weight to wells closest to the maternal residence. For example, an IDW well count of 125 wells/mile could be computed from 125 wells each located 1 mile from the maternal residence or 25 wells each located 0.2 miles from the maternal residence. We calculated the IDW well count of all existing natural gas wells in the birth year within a 10-mile radius of each maternal residence as a continuous exposure metric:

$$\text{IDW well count} = \sum_{i=1}^{n} \frac{1}{d_i} \qquad (1)$$

where IDW well count is the IDW count of existing wells within a 10-mile radius of maternal residence in the birth year; d_i is the distance of the ith individual well from maternal residence; and n is the number of existing wells within a 10-mile radius of maternal residence in the birth year.

The IDW well count was calculated for each maternal residence with ≥ 1 gas wells within 10 miles. The final distribution then was divided into tertiles (low, medium, and high) for subsequent logistic and linear regression analysis. Each tertile was compared with the referent group (no natural gas wells within 10 miles, IDW well count = 0).

Statistical analysis. We used logistic regressions to study associations between each dichotomous outcome and IDW exposure group. We also considered term birth weight as a continuous outcome using multiple linear regression. First, we estimated the crude odds ratio (OR) associated with IDW exposure tertile for each binary outcome, followed by a Cochran–Armitage test to evaluate linear trends in binominal proportions with increasing IDW exposure (none, low, medium, and high). We further investigated associations by adjusting for potential confounders, as well as infant and maternal covariates selected based on both a priori knowledge and empirical consideration of their association with exposure and an outcome. Specifically, covariates in our analysis of all outcomes except outcomes with very few events (i.e., NTDs, conotruncal defects, EMDs, and TVDs) included maternal age, education (< 12, 12, 13–15, ≥ 16 years), tobacco use (smoker, nonsmoker), ethnicity (Hispanic, non-Hispanic white), and alcohol use (yes, no), as well as parity at time of pregnancy (0, 1, 2, > 2) and infant sex. Gestational age was also included in the analysis of term birth weight. Elevation of maternal residence also was considered in the analysis because most wells are < 7,000 feet, and elevation has been associated with both preterm birth and low birth weight (Niermeyer et al. 2009). For 272 births where elevation of maternal residence was missing, elevation was imputed using mean elevation for maternal ZIP code. For outcomes with very few events, only elevation was included in the multiple logistic modeling to avoid unstable estimates. The ORs and their 95% CIs were used to approximate relative risks for each outcome associated with IDW count exposure tertile (low, medium, and high) compared with no wells within 10 miles, which is reasonable because of the rarity of the outcomes. We considered the statistical significance of the association, as well as the trend, in evaluating results, at an alpha of 0.05. We evaluated the confounding potential of the 1998 introduction of folic acid fortification on the birth defect outcomes and found only a decrease in NTD prevalence after 1998 (see Supplemental Material, Table S1).

In a sensitivity analyses, we explored reducing exposure to 2- and 5-mile buffers around the maternal residence, as well as restricting the cohort to births occurring between 2000 and 2009 to exclude births before the expansion of NGD. We report estimated associations with 95% CIs. All statistical analyses were conducted using SAS® software version 9.3 (SAS Institute Inc., Cary, NC).

TABLE 1: Study population characteristics for unexposed and exposed subjects from rural Colorado 1996–2009.

Maternal or infant charac-teristic	Total	Referent group (0 wells within 10 miles)	Low (first tertile)[a]	Medium (second tertile)[a]	High (third tertile)[a]
Total n (%)	124,842	66,626 (53)	19,214 (15)	19,209 (15)	19,793 (16)
Median	27	27	26	27	27
25th percentile	22	22	21	22	23
75th percentile	32	32	30	31	31
Maternal ethnicity (%)[b]					
Non-Hispanic white	73	74	72	76	69
Sex (%)					
Male	51	51	51	51	51
Maternal smoking (%)[c]					
Smokers	11	11	14	13	8
Maternal alcohol (%)[c]					
No	99	98	99	99	99
Parity (%)					
0	33	33	31	32	32
1	23	23	24	24	25
2	19	19	20	19	20
> 2	25	25	26	25	24
Residential elevation (feet)					
Median	5,000–5,999	6,000–6,999	< 5,000	5,000–5,999	< 5,000
25th percentile	< 5,000	5,000–5,999	< 5,000	< 5,000	< 5,000
75th percentile	7,000–7,999	7,000–7,999	5,000–5,999	6,000–6,999	5,000–5,999
Maternal education (%)					
< 12 years	21	20	26	19	22
12 years	30	30	33	29	28
13–15 years	23	22	25	25	24
≥ 16 years	26	28	18	26	27

[a]First tertile, 1–3.62 wells/mile; second tertile, 3.63–125 wells/mile; third tertile, 126–1,400 wells/mile. [b]Includes both Non-Hispanic and Hispanic white. [c]During pregnancy.

TABLE 2: Association between inverse distance weighted well count within 10-mile radius of maternal residence and CHDs, NTDs, and oral clefts.

Inverse distance weighted well count[a]	0 wells within 10 miles	Low (first tertile)	Medium (second tertile)	High (third tertile)	Cochran–Armitage trend test p-value[b]
Live births (n)	66,626	19,214	19.209	19,793	
CHDs					
Cases (n)	887	281	300	355	
Crude OR	1	1.1	1.2	1.3	< 0.0001
Adjusted OR (95% CI)[c]		1.1 (0.93, 1.3)	1.2 (1.0, 1.3)	1.3 (1.2, 1.5)	
NTDs					
Cases (n)	27	6	7	19	
Crude OR	1	0.77	0.90	2.4	0.01
Adjusted OR (95% CI)[d]		0.65 (0.25, 1.7)	0.80 (0.34, 1.9)	2.0 (1.0, 3.9)	
Oral clefts					
Cases (n)	139	31	41	40	
Crude OR	1	0.77	1	0.97	0.9
Adjusted OR (95% CI)[c]		0.65 (0.43, 0.98)	0.89 (0.61, 1.3)	0.82 (0.55, 1.2)	

[a]*First tertile, 1–3.62 wells/mile; second tertile, 3.63–125 wells/mile; third tertile, 126–1,400 wells/mile.* [b]*Performed as two-tailed test on unadjusted logistic regression.* [c]*Adjusted for maternal age, ethnicity, smoking, alcohol use, education, and elevation of residence, as well as infant parity and sex.* [d]*Adjusted only for residence elevation because of low numbers.*

7.3 RESULTS

Births were approximately evenly divided between exposed and unexposed groups (0 wells in a 10-mile radius versus ≥ 1 well in a 10-mile radius) (Table 1). Estimated exposure, represented by IDW well counts, tended to be higher for births to mothers with residence addresses at lower elevations (< 6,000 feet), and among nonsmoking and Hispanic mothers (Table 1).

Both crude and adjusted estimates indicate a monotonic increase in the prevalence of CHDs with increasing exposure to NGD, as represented by IDW well counts (Table 2). Births to mothers in the most exposed tertile (> 125 wells/mile) had a 30% greater prevalence of CHDs (95% CI: 1.2, 1.5) than births to mothers with no wells within a 10-mile radius of their residence.

Prevalence of NTDs was positively associated with only the third exposure tertile, based on crude and estimated adjusted ORs for elevation (Table 2). Births in the highest tertile (> 125 wells/mile) were 2.0 (95% CI: 1.0, 3.9) times more likely to have a NTD than those with no wells within a 10-mile radius, based on 59 available cases. We observed no statistically significant associations between oral clefts and NGD, based on trend analysis across categorical IDW well count exposure (Table 2).

TABLE 3: Association between inverse distance weighted well count within 10-mile radius of maternal residence and preterm birth and term low birth weight.

Inverse distance weighted well count[a]	0 wells within 10 miles	Low (first tertile)	Medium (second tertile)	High (third tertile)	Cochran–Armitage trend test p-value[b]
Preterm birth					
Live births (n)	65,506	18,884	18,854	19,384	
Cases (n)	4,849	1,358	1,289	1,274	
Crude OR	1	0.97	0.92	0.88	< 0.0001
Adjusted OR (95% CI)[c]		0.96 (0.89, 1.0)	0.93 (0.87, 1.0)	0.91 (0.85, 0.98)	
Term low birth weight					
Full-term live births (n)	60,653	17,525	17,565	18,104	
Cases (n)	2,287	525	471	432	
Crude OR	1	0.79	0.70	0.62	< 0.0001
Adjusted OR (95% CI)[c]		1.0 (0.9, 1.1)	0.86 (0.77, 0.95)	0.9 (0.8, 1)	
Mean difference in birth weight (g)[d]	0	5 (–2.2, 13)	24 (17, 31)	22 (15, 29)	

[a]First tertile, 1–3.62 wells/mile; second tertile, 3.63–125 wells/mile; third tertile, 126–1,400 wells/mile. [b]Performed as two-tailed test on unadjusted logistic regression. [c]Adjusted for maternal age, ethnicity, smoking, alcohol use, education, and elevation of residence, as well as infant parity and sex. [d]Adjusted for maternal age, ethnicity, smoking, alcohol use, education, and elevation of residence, as well as infant parity, sex, and gestational age.

TABLE 4: Association between inverse distance weighted well count within 10-mile radius of maternal residence and CHD diagnostic groups.

Inverse distance weighted well count[a]	0 wells within 10 miles	Low (first tertile)	Medium (second tertile)	High (third tertile)
Conotruncal defects				
Cases (n)	40	14	13	15
Adjusted OR (95% CI)[b]	1	1.1 (0.57, 2.2)	1.1 (0.55, 2.0)	1.2 (0.6, 2.2)
Ventricular septal defects				
Cases (n)	210	68	59	84
Adjusted OR (95% CI)[c]	1	1.3 (0.96, 1.8)	1.1 (0.81, 1.5)	1.5 (1.1, 2.1)
Endocardial cushion and mitrovalve defects				
Cases (n)	39	14	12	12
Adjusted OR (95% CI)[b]		0.81 (0.42, 1.6)	0.80 (0.41, 1.5)	0.67 (0.33, 1.32)
Pulmonary artery and valve defects				
Cases (n)	137	52	62	66
Adjusted OR (95% CI)[c]	1	1.3 (0.89, 1.8)	1.5 (1.1, 2,1)	1.6 (1.1, 2,2)
Tricuspid valve defects				
Cases (n)	9	5	8	8
Adjusted OR (95% CI)[b]	1	2.6 (0.75, 9.1)	3.9 (1.3, 11)	4.2 (1.3, 13)
Aortic artery and valve defects				
Cases (n)	75	22	21	24
Adjusted OR (95% CI)[c]	1	1.1 (0.68, 1.9)	1.0 (0.62, 1.8)	1.2 (0.73, 2.1)
Patent ductus arteriosis				
Cases (n)	59	18	17	15
Adjusted OR (95% CI)[c]	1	1.0 (0.56, 1.8)	0.96 (0.55, 1.7)	0.83 (0.44, 1.5)

[a]First tertile, 1–3.62 wells/mile; second tertile, 3.63–125 wells/mile; third tertile, 126–1,400 wells/mile. [b]Adjusted only for residence elevation of because of low numbers. [c]Adjusted for maternal age, ethnicity, smoking, alcohol use, education, and elevation of residence, as well as infant parity and sex.

Both crude and adjusted estimates for preterm birth suggest a slight (< 10%) decreased risk of preterm birth with increasing exposure to NGD (Table 3). Crude term low birth weight measures suggested decreased risk of term low birth weight with increasing exposure to NGD. A weak nonlinear trend remained after adjusting for elevation and other covariates. This association is consistent with the multiple linear regression results for continuous term birth weight, in which mean birth weights were 5–24 g greater in the higher IDW well count exposure tertiles than the referent group.

We observed a monotonic increase in the prevalence of NTDs with increasing exposure to NGD in our sensitivity analyses using 2- and 5-mile exposure radii as well as some attenuation in decreased risk for preterm birth and term low birth weight (see Supplemental Material, Tables S2–7). Restricting births to 2000 through 2009, the period of most intense NGD in Colorado, attenuated the positive association between NTDs in the highest tertile and did not alter observed relationships for other birth outcomes (see Supplemental Material, Tables S2–S7).

Exploratory analysis of CHDs by clinical diagnostic groups indicates increased prevalence of PAV defects by 60% (95% CI: 1.1, 2.2), VSDs by 50% (95% CI: 1.1, 2.1), and TVDs by 400% (95% CI: 1.3, 13) in the most exposed tertile compared with those with no wells within a 10-mile radius (Table 4).

7.4 DISCUSSION

We found positive associations between density and proximity of natural gas wells within a 10-mile radius of maternal residence and birth prevalence of CHDs and possibly NTDs. Prevalence of CHDs increased monotonically from the lowest to highest exposure tertile, although even in the highest tertile the magnitude of the association was modest. Prevalence of NTDs was elevated only in the highest tertile of exposure. We also observed small negative associations between density and proximity of natural gas wells within a 10-mile radius of maternal residence and preterm birth and term low birth weight, and a small positive association with mean birth weight. We found no indication of an association between density and proximity of natural gas wells within a 10-mile radius of maternal residence and oral cleft prevalence.

Nongenetic risk factors for CHDs and NTDs possibly attributable to NGD include maternal exposure to benzene (Lupo et al. 2010b; Wennborg et al. 2005), PAHs (Ren et al. 2011), solvents (Brender et al. 2002;

Chevrier et al. 1996; Desrosiers et al. 2012; McMartin et al. 1998), and air pollutants (NO2, SO2, PM) (Vrijheid et al. 2011). NGD emits multiple air pollutants, including benzene and toluene, during the "well completion" phase (when gas and water flow back to the surface after hydraulic fracturing) as well as from related infrastructure (CDPHE 2009a, 2009b; Garfield County Public Health Department 2009; Gilman et al. 2013; McKenzie et al. 2012; Pétron et al. 2012). Ambient benzene levels in areas with active NGD in Northeast Colorado ranged from 0.03 to 6 parts per billion by volume (ppbv) (CDPHE 2012; Gilman et al. 2013; Pétron et al. 2012). Furthermore, 24-hr average ambient air benzene levels near active well development sites in western Colorado ranged from 0.03 to 22 ppbv (McKenzie et al. 2012).

Two previous case–control studies have reported associations between maternal exposure to benzene and birth prevalence of NTDs and/or CHDs (Lupo et al. 2010b; Wennborg et al. 2005). The study by Lupo et al. (2010b) of 4,531 births in Texas found that mothers living in census tracts with the highest ambient benzene levels (0.9–2.33 ppbv) were 2.3 times more likely to have offspring with spina bifida than mothers living in census tracts with the lowest ambient benzene levels (95% CI: 1.22, 4.33). An occupational study of Swedish laboratory employees found a significant association between exposure to occupational levels of benzene in the critical window between conception, organogenesis, and neural crest formation and neural crest malformations (Wennborg et al. 2005). Children born to 298 mothers exposed to benzene had 5.3 times greater prevalence of neural crest malformations than children born to mothers not exposed to benzene (95% CI: 1.4, 21.1). Other studies of maternal exposures to organic solvents, some of which contain benzene, have reported associations between maternal occupational exposure to organic solvents and major birth defects (Brender et al. 2002; Desrosiers et al. 2012; McMartin et al. 1998). Although exposure to benzene is a plausible explanation for the observed associations, further research is needed to examine whether these associations are replicated and whether benzene specifically explains these associations.

Air pollutants emitted from diesel engines used extensively in NGD also may be associated with CHDs and/or NTDs. Trucks with diesel engines are used to transport supplies, water, and waste to and from gas wells, with 40 to 280 truck trips per day per well pad during development (Witter et al. 2013). Generators equipped with diesel engines are used in both drilling wells and hydraulic fracturing. Air pollutants in diesel exhaust include NO_2, SO_2, PM, and PAHs. A meta-analysis of four studies suggested associations of maternal NO_2 and SO_2 exposures with coarctation of the aorta and tetralogy of Fallot, and of maternal PM_{10} exposure with arterial septal defects (Vrijheid et al. 2011). Two case–control studies in China reported positive associations between PAH concentrations in maternal blood and the placenta and NTDs (Li et al. 2011; Naufal et al. 2010). Several CHDs were associated with traffic related carbon monoxide and ozone pollution in a case control study of births from 1987 to 1993 in Southern California (Ritz et al. 2002).

The small negative associations with term low birth weight and preterm birth in our study population were unexpected given that other studies have reported postive associations between these outcomes and urban air pollution (Ballester et al. 2010; Brauer et al. 2008; Dadvand et al. 2013; Ghosh et al. 2012; Llop et al. 2010) and proximity to natural gas wells (Hill 2013). It is possible that rural air quality near natural gas wells in Colorado is not as compromised as urban air quality in these studies, and exposure represented as IDW well count may not adequately represent air quality. In addition, the power of our large cohort increases the likelihood of false positive results for small associations close to the null. Although associations were consistent across measures of birth weight (i.e., reduced risk of term low birth weight and increase in mean birth weight), they attenuated toward the null in sensitivity analysis for 2- and 5-mile radii (see Supplemental Material, Tables S6–S7). If causal, stronger associations would be expected with more stringent exposure definions. Our incomplete ability to adjust for socioeconomic status, health, nutrition, prenatal care, and pregnancy complications likely accounts for these unexpected findings.

This study has several limitations inherent in the nature of the available data. Not all birth defects were confirmed by medical record review. Also, birth defects are most likely undercounted, because stillbirths, terminated pregnancies, and later-life diagnoses (after 3 years of age) are not included. Birth weight and gestational age were obtained from birth certificates, which are generally accurate for birth weight and useful but less accurate for gestational age (DiGiuseppe et al. 2002). Data on co-variates were obtained from birth certificates and were limited to basic demographic, education, and behavioral information available in the vital records. Distribution of covariates among exposure tertiles and the unexposed group was similar; nevertheless, our incomplete ability to adjust for socioeconomic status, health, nutrition, prenatal care, and pregnancy complications may have resulted in residual confounding. In addition, low event outcomes (e.g., NTDs) were adjusted only for elevation. The data set did not contain information on maternal folate consumption and genetic anomalies, both independent predictors of our outcomes, which may have confounded these results. We did observe a large decrease in the prevalence of NTDs after the introduction of folic acid in 1998, and small increases in the prevalence of CHDs and oral clefts, although none of the estimates are statistically significant (see Supplemental Material, Table S1). Further study is needed to determine whether unaccounted folate confounding is attenuating our results toward the null. There is no evidence indicating genetic anomalies would differ by IDW well count around maternal residence.

Because of the rarity of specific birth defects in the study population, birth defects were aggregated into three general groups. This limited our study in that associations with specific birth defects may have been obscured. An exploratory analysis of CHDs by clinical diagnostic groups indicates increased prevalence of specific diagnostic groups (i.e., PAV, VSD, and TVD) compared with aggregated CHDs (Table 4).

Another limitation of this study is the lack of temporal and spatial specificity of the exposure assessment. Because we did not have maternal residential history, we assumed that maternal address at time of delivery was the same as maternal address during the first trimester of pregnancy—the critical time period for formation of birth defects. Studies in Georgia and Texas estimate that 22–30% of mothers move residence during their pregnancy, and most mothers move within their locality (Lupo et al. 2010a; Miller et al. 2010), potentially introducing some exposure misclassification for the early pregnancy period of interest. However, these studies found little difference in mobility between cases and controls (Lupo et al. 2010a; Miller et al. 2010), and maternal mobility did not significantly influence the assessment of benzene exposure (Lupo et al. 2010a). We were able to determine only whether a well existed within the calendar year of birth (e.g., 2003) and did not have sufficient data to determine if a well existed within the first trimester of the pregnancy. Therefore, some nondifferential exposure misclassification is likely and the overall effect of this is unknown.

Similarly, we had consistent information only on existence of a well in the birth year. Lack of information on natural gas well activity levels, such as whether or not wells were producing or undergoing development, may have resulted in exposure misclassification. Actual exposure to natural gas–related pollutants likely varies by intensity of development activities. Lack of temporal and spatial specificity of the exposure assessment would most likely have tended to weaken associations (Ritz et al. 2007; Ritz and Wilhelm 2008). To address spatial and temporal variability, additional air pollution measurements and modeling will be needed to improve exposure estimates at specific locations. Last, information on the mother's activities away from her residence, such as work and recreation, as well as proximity of these activities to NGD was not available and may have led to further exposure misclassification and residual confounding.

7.5 CONCLUSION

This study suggests a positive association between greater density and proximity of natural gas wells within a 10-mile radius of maternal residence and greater prevalence of CHDs and possibly NTDs, but not oral clefts, preterm birth, or reduced fetal growth. Further studies incorporating information on specific activities and production levels near homes over the course of pregnancy would improve exposure assessments and provide more refined effect estimates. Recent data indicate that exposure to NGD activities is increasingly common. The COGCC estimates that 26% of the > 47,000 oil and gas wells in Colorado are located within 150–1,000 feet of a home or other type of building intended for human occupancy (COGCC 2012). Taken together, our results and current trends in NGD underscore the importance of conducting more comprehensive and rigorous research on the potential health effects of NGD.

REFERENCES

1. Agency for Toxic Substances and Disease Registry. 2007. Toxicological Profile for Benzene. Available: http://www.atsdr.cdc.gov/ToxProfiles/TP.asp?id=40&tid=14 [accessed 22 May 2013].
2. Ballester F, Estarlich M, Iniguez C, Llop S, Ramon R, Esplugues A, et al. 2010. Air pollution exposure during pregnancy and reduced birth size: a prospective birth cohort study in Valencia, Spain. Environ Health 9:6; doi: 10.1186/1476-069X-9-6.
3. Bowen S, Irtenkauf S, Hannigna J, Stefanski A. 2009. Alterations in rat morphology following abuse patterns of toluene exposure. Reprod Toxicol 27:161–169.
4. Brauer M, Lencar C, Tamburic L, Koehoorn M, Demers P, Karr C. 2008. A cohort study of traffic-related air pollution impacts on birth outcomes. Environ Sci Technol 116:680–686.
5. Brender JD, Suarez L, Hendricks KA, Baetz RA, Larsen R. 2002. Parental occupation and neural tube defect-affected pregnancies among Mexican Americans. J Occup Environ Med 44:650–656.
6. Brent RL. 2004. Environmental causes of human congenital malformations: the pediatrician's role in dealing with these complex clinical problems caused by a multiplicity of environmental and genetic factors. Pediatrics 113:957–968.
7. Bukowski JA. 2001. Review of the epidemiological evidence relating toluene to reproductive outcomes. Regul Toxicol and Pharmacol 33:147–156.
8. CDPHE (Colorado Department of Public Health and Environment). 2007. Garfield County Air Toxics Inhalation: Screening Level Human Health Risk Assessment: Inhalation Of Volatile Organic Compounds Measured In Rural, Urban, and Oil & Gas Areas In Air Monitoring Study (June 2005–May 2007). Available: http://www.garfield-county.com/public-health/documents/Working%20Draft%20CDPHE%20Screeing%20Level%20Risk%20Air%20Toxics%20Assessment%2012%20 20%2007.pdf [accessed 22 May 2013].
9. CDPHE (Colorado Department of Public Health and Environment). 2009a. Garfield County Emissions Inventory. Available: http://www.garfield-county.com/air-quality/documents/airquality/Garfield_County_Emissions_Inventory-2009.pdf [accessed 22 May 2013].
10. CDPHE (Colorado Department of Public Health and Environment). 2009b. State of Colorado Technical Support Document for Recommended 8-Hour Ozone Designations. Available: http://www.colorado.gov/cs/Satellite?blobcol=urldata&blobheadername1=Content-Disposition&blobheadername2=Content-Type&blobheadervalue1=inline%3B+filename%3D%22Recommended+8+Hour+Ozone+Area+Designations.pdf%22&blobheadervalue2=application%2Fpdf&blobkey=id&blobtable=MungoBlobs&blobwhere=1251808873080&ssbinary=true [accessed 22 May 2013].
11. CDPHE (Colorado Department of Public Health and Environment). 2012. Air Emissions Case Study Related to Oil and Gas Development in Erie, Colorado. Available: http://www.colorado.gov/airquality/tech_doc_repository.aspx [accessed 22 May 2013].
12. Centers for Disease Control and Prevention. 2013. Birth Defects. Available: http://www.cdc.gov/ncbddd/birthdefects/index.html [accessed 22 May 2013].

13. Chevrier C, Dananche B, Bahuau M, Nelva A, Herman C, Francannet C, et al. 1996. Occupational expsoure to organic solvent mixtures during pregancy and the the risk of non-syndromic oral clefts. Occup Environ Med 63:617–623.
14. COGCC (Colorado Oil and Gas Conservation Commission). 2009. Statement of Basis, Specific Statutory Authority, and Purpose: New Rules and Amendments to Current Rules of the Colorado Oil and Gas Conservation Commission, 2 ccr 404-1. Available: http://cogcc.state.co.us/ [accessed 22 May 2013].
15. COGCC (Colorado Oil and Gas Conservation Commission). 2012. Staff Report. Colorado Department of Natural Resources. Available: http://cogcc.state.co.us/ [accessed 22 May 2013].
16. COGIS (Colorado Oil and Gas Information System). 2011. Well Production Database. Vol. 2011. Available: http://cogcc.state.co.us/ [accessed 22 May 2013].
17. Colborn T, Kwiatkowski C, Schultz K, Bachran M. 2011. Natural gas operations from a public health perspective. Hum Ecol Risk Asses 17:1039–1056.
18. CRCSN (Colorado Responds to Children with Special Needs). 2011. Birth Defects Dataset Details. Available: http:// www.cdphe.state.co.us/cohid/crcsndata.html [accessed 6 October 2011].
19. Dadvand P, Parker JD, Bell ML, Bonzini M, Brauer M, Darrow L, et al. 2013. Maternal exposure to particulate air pollution and term birth weight: a multi-country evaluation of effect and heterogeneity. Environ Health Perspect 121:367–373; doi: 10.1289/ehp.1205575.
20. Desrosiers TA, Lawson CC, Meyer RE, Richardson DB, Daniels JL, Waters MA, et al. 2012. Maternal occupational exposure to organic solvents during early pregnancy and risks of neural tube defects and orofacial clefts. Occup Environ Med 69:493–499.
21. DiGiuseppe DL, Aron DC, Ranbom L, Harper DL, Rosenthal GE. 2002. Reliability of birth certificate data: a multi-hospital comparison to medical records information. Matern Child Health J 6:169–179.
22. Esswein EJ, Breitenstein M, Snawder J, Kiefer M, Sieber K. 2013. Occupational exposure to respirable crystalline silica during hydraulic fracturing. J Occup Environ Hyg 10(7):347–356; doi: 10.1080/15459624.2103.788352.
23. Frazier A. 2009. Analysis of Data Obtained for the Garfield County Air Toxics Study—Summer 2008. Available: http://www.garfield-county.com/air-quality/documents/airquality/2008_Targeted_Oil_and_Gas_Monitoring_Report.pdf   [accessed 22 May 2013].
24. Garfield County Public Health Department. 2009. Garfield County 2009 Air Quality Monitoring Summary. Available: http://www.garfield-county.com/air-quality/documents/airquality/2009_Air_Monitoring_Report.pdf [accessed 22 May 2013].
25. Ghosh JKC, Wilhelm M, Su J, Goldberg D, Cockburn M, Jerrett M, et al. 2012. Assessing the influence of traffic-related air pollution on risk of term low birth weight on the basis of land-use-based regression models and measures of air toxics. Am J Epidemiol 175:1262–1274.
26. Gilboa SM, Mendola P, Olshan AF, Langlois PH, Savitz DA, Loomis D, et al. 2005. Relation between ambient air quality and selected birth defects, Seven County Study, Texas, 1997–2000. Am J Epidemiol 162(3):238–252.

27. Gilman JB, Lerner BM, Kuster WC, de Gouw J. 2013. Source signature of volatile organic compounds (VOCs) from oil and natural gas operations in northeastern Colorado. Environ Sci Technol 47(3):1297–1305.

28. Hill EL. 2013. Unconventional Natural Gas Development and Infant Health: Evidence from Pennsylvania. Ithaca, NY:Charles Dyson School of Applied Economics and Management, Cornell University. Available: http://dyson.cornell.edu/research/researchpdf/wp/2012/Cornell-Dyson-wp1212.pdf [accessed 7 March 2014].

29. Honein M, Rasmussen S, Reefhuis J, Romitti P, Lammer E, Sun L, et al. 2006. Maternal smoking and environmental tobacco smoke exposure and the risk of orofacial clefts. Epidemiology 18:226–233.

30. Kemball-Cook S, Bar-Ilan A, Grant J, Parker L, Jung J, Santamaria W, et al. 2010. Ozone impacts of natural gas development in the Haynesville shale. Environ Sci Technol 44:9357–9363.

31. King GE. 2012. Hydraulic Fracturing 101: What Every Representative, Environmentalist, Regulator, Reporter, Investor, University Researcher, Neighbor and Engineer Should Know About Estimating Frac Risk and Improving Frac Performance in Unconventional Gas and Oil Wells. Available: http://fracfocus.org/sites/default/files/publications/hydraulic_fracturing_101.pdf [accessed 6 March 2014].

32. Li Z, Zhang L, Ye R, Pei L, Liu J, Zheng X, et al. 2011. Indoor air pollution from coal combustion and the risk of neural tube defects in a rural population in Shanxi Province, China. Am J Epidemiol 174:451–458.

33. Llop S, Ballester F, Estarlich M, Esplugues A, Rebagliato M, Iñiguez C. 2010. Preterm birth and exposure to air pollutants during pregnancy. Environ Res 110:778–785.

34. Lupo PJ, Symanski E, Chan W, Mitchell LE, Waller DK, Canfield MA, et al. 2010a. Differences in exposure assignment between conception and delivery: the impact of maternal mobility. Paediatr Perinat Epidemiol 24:200–208.

35. Lupo P, Symanski E, Waller D, Chan W, Langlosi P, Canfield M, et al. 2010b. Maternal exposure to ambient levels of benzene and neural tube defects among offspring, Texas 1999–2004. Environ Health Perspect 119:397–402; doi: 10.1289/ehp.1002212.

36. McKenna MT, Michaud CM, Murray CJL, Marks JS. 2005. Assessing the burden of disease in the United States using disability-adjusted life years. Am J Prev Med 28:415–423.

37. McKenzie LM, Witter RZ, Newman LS, Adgate JL. 2012. Human health risk assessment of air emissions from development of unconventional natural gas resources. Sci Total Environ 424:79–87.

38. McMartin KI, Chu M, Kopecky E, Einarson TR, Koren G. 1998. Pregnancy outcome following maternal organic solvent exposure: a meta-analysis of epidemiologic studies. Amn J Ind Med 34:288–292.

39. Miller A, Siffel C, Correa A. 2010. Residential mobility during pregnancy: patterns and correlates. Matern Child Heatlh J 14:625–634.

40. National Center for Health Statistics. 2011. Classification of Diseases and Injuries. Available: ftp://ftp.cdc.gov/pub/Health_Statistics/NCHS/Publications/ICD-9/ucod.txt [accessed 22 May 2013].

41. Naufal Z, Zhiwen L, Zhu L, Zhou GOD, McDonald T, He LY, et al. 2010. Biomarkers of exposure to combustion by-products in a human population in Shanxi, China. J Expos Sci Environ Epidemiol 20:310–319.

42. Niermeyer S, Andrade Mollinedo P, Huicho L. 2009. Child health and living at high altitude. Arch Dis Child 94:806–811.

43. Olaguer EP. 2012. The potential near-source ozone impacts of upstream oil and gas industry emissions. J Air Waste Manag Assoc 62:966–977.

44. Parker SE, Mai CT, Canfield MA, Rickard R, Wang Y, Meyer RE, et al. 2010. Updated national birth prevalence estimates for selected birth defects in the United States, 2004–2006. Birth Defects Res A Clin Mol Teratol 88:1008–1016.

45. Pétron G, Frost G, Miller BR, Hirsch AI, Montzka SA, Karion A, et al. 2012. Hydrocarbon emissions characterization in the Colorado front range: a pilot study. J Geophys Res 117:D04304; doi: 10.1029/2011JD016360.

46. Ren A, Qiu X, Jin L, Ma J, Li Z, Zhang L, et al. 2011. Association of selected persistent organic pollutants in the placenta with the risk of neural tube defects. Proc Natl Acad Sci USA 108:12770–12775.

47. Ritz B, Wilhelm M. 2008. Ambient air pollution and adverse birth outcomes: Methodologic issues in an emerging field. Basic Clinl Pharmacol Toxicol 102:182–190.

48. Ritz B, Wilhelm M, Hoggatt KJ, Ghosh JKC. 2007. Ambient air pollution and preterm birth in the environment and pregnancy outcomes study at the University of California, Los Angeles. Am J of Epidemiol 166:1045–1052.

49. Ritz B, Yu F, Fruin S, Chapa G, Shaw GM, Harris JA. 2002. Ambient air pollution and risk of birth defects in southern California. Am J Epidemiol 155:17–25.

50. Romitti P, Sun L, Honein M, Reefhuis J, Correa A, Rasmussen S. 2007. Maternal periconceptual alcohol consumption and risk of orofacial clefts. Am J Epidemiol 166:775–785.

51. Shepard T. 1995. Agents that cause birth defects. Yonsei Med J 36:393–396.

52. U.S. EIA (U.S. Energy Information System). 2011a. International Energy Outlook 2011. DOE/EIA-0484(2011). Washington, DC:U.S. EIA.

53. U.S. EIA (U.S. Energy Information System). 2011b. Review of Emerging Resources: U.S. Shale Gas and Shale Oil Plays. Available: http://www.eia.gov/analysis/studies/usshalegas/pdf/usshaleplays.pdf [accessed 22 May 2013].

54. Vidas H, Hugman B. 2008. Availability, Economics, and Production Potential of North American Unconventional Natural Gas Supplies. Available: http://www.ingaa.org/File.aspx?id=7878 [accessed 22 May 2013].

55. Vrijheid M, Martinez D, Manzanares S, Dadvand P, Schembari A, Rankin J, et al. 2011. Ambient air pollution and risk of congenital anomalies: a systematic review and meta-analysis. Environ Health Perspect 119:598–606; doi: 10.1289/ehp.1002946.

56. Wald N, Sneddon J. 1991. Prevention of neural tube defects: Results of the medical research council vitamin study. Lancet 338:131.

57. Walther E. 2011. Screening Health Risk Assessment Sublette County, Wyoming. SR2011-01-03. Available: http://www.sublettewyo.com/DocumentCenter/Home/View/438 [accessed 22 May 2013].

58. Wennborg H, Magnusson L, Bonde J, Olsen J. 2005. Congenital malformations related to maternal exposure to specific agents in biomedical research laboratories. J Occup Environ Med 47:11–19.

59. Witter R, McKenzie L, Stinson K, Scott K, Newman L, Adgate J. 2013. The use of health impact assessment for a community undergoing natural gas development. Am J Public Health 103(6): doi: 10.2105/AJPH.2012.301017.
60. Zielinska B, Fujita E, Campbell D. 2011. Monitoring of Emissions from Barnett Shale Natural Gas Production Facilities for Population Exposure Assessment. Desert Research Institute. Available: https://sph.uth.edu/mleland/attachments/Barnett%20 Shale%20Study%20Final%20Report.pdf [accessed 22 May 2013].

There are several supplemental files that are not available in this version of the article. To view this additional information, please use the citation on the first page of this chapter.

PART IV

LEAD EXPOSURE

CHAPTER 8

EXPLORING CHILDHOOD LEAD EXPOSURE THROUGH GIS: A REVIEW OF THE RECENT LITERATURE

CEM AKKUS AND ESRA OZDENEROL

8.1 INTRODUCTION

The use of GIS in environmental risk factor studies on childhood lead exposure became a focus of research activity in the late 1990s. This prompted the CDC to develop a guideline for the use of GIS in childhood lead poisoning studies in 2004 [1]. Even though the number of children with elevated blood lead levels (EBLLs) in the U.S. is decreasing, eliminating EBLLs by the year 2020 remains a goal of the U.S. Department of Health and Human Services [2]. The capacity to achieve this goal is conditional on the ability to develop strategies based on geographic areas [3]. Funding is another factor to achieve this goal especially when health departments have limited budgets [4]. Despite significant research on the risk factors affecting childhood lead poisoning (age of housing, urban/rural status, race/ethnicity, socioeconomic status, population density, renter/owner occupancy, housing value, nutritional status), there has not been any review article discussing the GIS-based studies. The purpose of this article is to

Exploring Childhood Lead Exposure through GIS: A Review of the Recent Literature. © *Akkus C and Ozdenerol E.* International Journal of Environmental Research and Public Health *11,6 (2014), doi: 10.3390/ijerph110606314. Licensed under Creative Commons Attribution 3.0 Unported License, http://creativecommons.org/licenses/by/3.0/.*

review previous and current GIS research to understand which methods currently employed have been most effective in the screening strategies and examining spatial epidemiology of childhood lead exposure. Another goal is to identify additional methods in GIS-utilized lead poisoning research that also provide public health practitioners and policy makers the ability to better target lead poisoning preventive interventions. Our review covers the time period from 1991 to 2012 and includes GIS-based studies which were published until the adoption of the toxicity threshold of blood lead levels of 5 microgram per-deciliter (μg/dL) by the CDC [5].

8.1.1 ECOLOGICAL STUDIES AND GIS USE IN CHILDHOOD LEAD POISONING

Ecological studies focusing on the distribution of blood lead levels, susceptible populations, and exposure sources have been cited to address childhood lead exposure. Identification of environmental risk factors and understanding of the distribution of the lead in the environment is important for health departments in better targeting at risk populations [6,7,8,9]. Ecological studies modeling risk factors are also valuable because they give insight to public health intervention strategies [10,11,12,13,14,15,16,17,18]. For ecological studies of childhood lead poisoning, one needs to identify sources of lead toxicity and determine environmental risk factors based on the distribution of the toxicants and how children come into contact with them in their daily lives [19,20,21,22,23,24,25,26,27,28]. Children's bodies absorb lead easily, especially in the brain and central nervous system, making them highly susceptible to the effects of lead poisoning. Sources of environmental lead contamination can be difficult to pinpoint because the pathways to lead absorption are various: (1) deteriorating lead-based paint from walls, windows, and doors; (2) transportation of lead contamination to the house by other means; (3) playing with toys which contain lead; (4) absorption of leaded dust through hand-to-mouth behaivour; and (5) being in polluted environment [29]. The most common pathway could be hand-to-mouth behavior especially among young children, however it is hard to know when and how they interact with lead contamination [30]. Exposure during childhood is thought to be brief, usually until the age of 6

[31]; however, the side effects persist throughout life [32]. Possible sources for lead include: leaded paint, lead contaminated soil, lead in plumbing, automobile exhaust, by-products of both mining and metal working, and various consumer products [18,33,34,35,36]. After the ill effects of lead on people's health were recognized, lead was first banned in Europe in the early 1900s [37]. Lead use in the US was successively banned in paint (1978) [38], in pipes (1986) [39], and in gasoline (1995) [40]. Environmental lead from these sources has not been completely eliminated. Houses with old pipes and paint, which contaminate the drinking water and surrounding soil, are still a significant source of lead exposure [31,36,41].

Despite being a preventable environmental problem, lead poisoning remains a major health threat and a persistent source of illness in the United States. Its estimated cost is $50.9 billion [42]. Changes in federal laws to limit the use of lead reversed the increasing trend in BLLs of children in the US between 1900 and 1975, but children aged <6 years continued to be exposed to lead [31]. In the US, the threshold of elevated blood lead level (EBLL) for childhood lead poisoning has changed four times over the last four decades. Before 1975, lead concentrations of 60 µg/dL and above were considered elevated. With our increased understanding of lead poisoning, the threshold has lowered to 30 µg/dL in 1975, 25 µg/dL in 1985, 10 µg/dL in 1991, and finally 5 µg/dL in 2012 [43,44,45,46,47]. To date, no safe blood lead thresholds for the adverse effects of lead on children have been identified [31]. GIS use in childhood lead poisoning studies started in the 1990s. In 1992, Wartenberg [48] conducted one of the earliest GIS studies on childhood lead poisoning by focusing on theoretical GIS methodologies rather than data analysis. Public health departments recognized the advantages of GIS in screening, exposure prediction, and mapping cases. Using BLL data for lead poisoning, an increasing number of GIS-based ecological studies have identified risk factors as socioeconomic status (SES) [9,10,11,12,13,14,15,16,17,18,21], year built of housing [7, 8,9,10,11,13,15,16,17,18,20,21,23,28], race [11,13,14,16,17,21,23,27,28] and ethnicity [15,16,18].

In lead poisoning studies, GIS was used in various stages from data preparation, to multivariate mapping of BLLs with their risk factors, to spatial and statistical analysis. At the data preparation stage, address geocoding is the most used tool to transfer tabular data sets, such as screened children addresses, into GIS [7,8,9,10,11,12,13,14,15,16,17,18,19,23]. Various GIS functions were used for multivariate mapping of BLLs and risk factors in a limited custom such as linking SES data with screened data records [49,50], map overlays [51,52], distance calculations [53], and hyperlinks to demolishing sites' photos and city maps for mapping dust-fall lead loadings [54]. More sophisticated spatial methods have also been used such as spatial clustering [15,18,21,24,26], spatial auto-correlation [10,13,15,18,21], spatial regression [14], and risk modeling [10,11,12,13,14,15,16,17,18]. New GIS-based studies are used in surveillance data management, risk analysis, lead exposure visualization, and community intervention strategies where geographically-targeted and specific intervention measures are taken.

8.1.2 RECENT REVIEWS

A review of GIS-utilized studies on childhood lead poisoning has not been conducted. There are some non-GIS based reviews on lead poisoning in relation to cardiovascular diseases [55], resuspension of urban soil [56], multiple risk factors on Hispanic sub-population [57], lead dust from traffic volume [58], leaded gasoline on urbanized areas [59], and exposure to lead in soil dust [60]. We will describe these reviews and summarize what is known and unknown as a source of lead exposure and build on these reviews with our comprehensive review, inclusive of GIS-based studies.

Navas-Acien et al. [55] studied lead exposure and cardiovascular disease in 2007. The authors reviewed studies regarding the association between BLLs and blood pressure, lead exposure and clinical cardiovascular disease in the general population, cardiovascular mortality in occupational populations exposed to lead, and lead exposure and intermediate cardiovascular end points. The review found a positive association but not a causal relationship between lead exposure and cardiovascular end points in general and occupational populations. The study also showed suggestive—but not causal—evidence that there is a relationship between lead exposure and heart rate variability. These associations were observed at low level BLLs (well below 5 µg/dL).

Laidlaw et al. published two reviews about the relationship between lead in soil and children blood lead levels in 2008 and 2011 [56,60]. In 2008, Laidlaw et al. [56] claimed that seasonality could be another source of lead poisoning problems besides paint chips, leaded soil, and pipes. Their review also discussed the study designs of "soil lead" vs. "blood lead" studies. They created a statistical model in order to investigate the atmospheric soil seasonality and the prediction model for atmospheric soil in the US. In terms of soil lead topology, they reviewed studies indicating that lead in soil decayed exponentially away from the historical main roads [61,62]. Another study by Mielke et al. [63] also suggested that changes in soil lead in the inner city might be better explained with historical lead deposits from traffic than from old housing (leaded paint). In their review in 2011, Laidlaw et al. [60] focused on Australian inner cities as they found that there were few studies conducted in the inner cities. The authors suggested that there should be high density soil lead mapping as well as universal screening in older neighborhoods in Australia's large inner cities.

Brown et al. [57] presented literature on sources of lead in Hispanic sub-populations which indicates children with Hispanic origin are at high risk in the population. The authors reviewed the literature for lead poisoning among Hispanic populations based on their location, behavior, and diet. In terms of location, the review suggested that there was a relationship between immigrant populations and lead poisoning. Among studies they reviewed, Cowan et al. found that children on the Mexican side of US-Mexico border had higher BLLs compared to the children who lived in the US side of the border. However, poverty could be a confounding factor in the area [57]. Another study [64] in their review showed that 43% of Mexican children had elevated BLLs ($\geq$10 μg/dL) in an area close to the border of El Paso, Texas. Location based studies include migratory farmworkers as well. Another location-dependent behavioral pathway was the consumption of lead glazed pottery bits known as "pica". This material was being consumed by women during their pregnancy in Mexico due to the belief that this material was helpful for the baby [65,66]. In terms of dietary intake, exposure to lead varies from folk remedies to imported candies. The review also suggested that there was food insecurity among Hispanic subpopulations which may result in iron deficiency which increases lead absorption in the bodies of children.

Mielke et al. published two reviews about environmental aspects of lead poisoning in consecutive years 2010 and 2011 [58,59]. In their 2010 review, Mielke et al. investigated the effect of traffic on lead poisoning regarding lead emissions and additives used in eight California urbanized areas. The authors used three datasets in order to show the gasoline lead contribution in the environment; annual lead amounts from 1927 to 1984, 1982 lead additive quantities for eight urbanized areas in California, and California fuel consumption data from 1950 to 1982. The review showed that there was a correlation between the lead amount in soil and size of the cities. Community location was also related to the lead amount. Inner cities where high traffic volume occurs had higher amounts of leaded soil compared to the suburbs. The review also showed that the distance decay characteristics of lead in soil were similar throughout the US. There was a strong correlation between children BLLs and lead in soil. Mielke's review confirmed the relationship between children BLLs and seasonality. Mielke et al. found a negative relationship between lead in soil and school erformance of children. In their second review in 2011, they expanded their previous California study to 90 urbanized areas throughout the US. Their findings corroborated the previous findings.

8.2 METHODS

A literature search was conducted to identify recent articles discussing childhood lead poisoning and the use of GIS and risk modeling. Several on-line databases were queried, including JSTOR, CINAHL, Web of Science, ScienceDirect, and PubMed. The following key words were used individually and in combination as inclusion criteria for articles to be considered for this review; children, childhood, pediatric, Pb, lead, poisoning, toxicity, geographic, information, systems, and GIS. Our review covers a 21 year period which includes GIS-based studies published since 10 µg/dL thresholds were first introduced in 1991 until the new threshold of 5 µg/dL in 2012. Initial searches yielded approximately 981 results. The abstracts of these papers were reviewed to confirm applicability. After considering additional exclusion criteria (manuscripts not having BLL data analysis, no GIS use, non-English language, and manuscripts not available as full-text), 23 papers remained.

TABLE 1: Summary of studies with common risk factors and major findings.

GIS Analysis/Citation	Region/Date	Common Risk Factors/Major Findings
Screening methodology design		
Overlay analysis, choropleth mapping/[6]	Knoxville, TN/1998	Old housing, and proximity to old roads/ The screening data based on the study's risk criteria thoroughly represents the targeted population.
Address geocoding, overlay analysis, choropleth mapping/[7]	Jefferson, KY/2001	Old housing/Percent children with EBLLs is strongly associated with old housing. The screening data based on the study's risk criteria does not fully represent the targeted population.
Address geocoding, overlay analysis/[8]	South Carolina/2003	Old housing/EBLLs are strongly associated with old housing. The screening data based on the study's risk criteria does not fully represent the targeted population.
Address geocoding, overlay analysis, choropleth mapping/[9]	Atlanta, GA/2009	Poverty, old housing/The screeing is strongly correlated with WIC (Special Supplemental Nutrition Program for Women, Infants and Children enrolment) status but not with old housing.
Risk modeling studies		
Spatial autocorrelation/ [10]	Rhode Island/1997	Old housing, poverty, vacancy, percent screened children, and percent immigrants/ Older houses and vacant housing are significantly associated with excessive childhood lead exposure.
Address geocoding, overlay analysis, choropleth mapping/[11]	Durham, NC/2002	Old housing, income, and race/The percentage of African American population, median income, and construction year of housings are significantly associated with childhood lead exposure.
Address geocoding/[12]	Rhode Island/2003	Poverty, education, occupation, wealth/BLLs are strongly associated with poverty but not education level, occupation, and wealth.
Spatial autocorrelation with Simultanious Autoregressive Model (SAR)/[13]	New York/2004	Old housing, race, poverty, population density, education, vacant housing, renting, and seasonality/The age of housing, education level, and percentage of African American population variables are significant predictors of BLLs.
Point in polygon analysis (PIP), address geocoding, and spatial regression/[14]	Syracuse, NY/2007	House value, race/EBLLs are significantly associated with the percentage of African American population and average house value.

TABLE 1: *Cont.*

GIS Analysis/Citation	Region/Date	Common Risk Factors/Major Findings
Spatial autocorrelation, kriging, Local Moran's I, and LISA/[15]	Cook, IL/2007	Old housing, income, and minority populations/The authors concluded that the dependent variable is significantly associated with housing age, income, and minority populations.
Address geocoding, risk modeling/[16]	North Carolina/2008	Old housing, race, percent Hispanic, income, poverty, and seasonality/All variables are significantly associated with childhood lead exposure.
Address geocoding, sensitivity analysis/[17]	Michigan/2010	Old housing, race, poverty, race, and education/BLL is associated with children's immediate environment than a larger area such as a census tract or ZIP code.
Spatial autocorrelation, kriging, Local Moran's I, and LISA/[18]	Cook, IL/2010	Old housing, income, and minority populations/The authors concluded that the dependent variable is significantly associated with housing age, income, and minority populations.
Environmental risk factors		
Address geocoding, choropleth mapping, and overlay analysis/ [19]	New Jersey/1992	Proximity to industrial sites emitting lead and hazardous waste sites contaminated with lead, and proximity to roads with high traffic volume.
3-D Surface Modeling/ [20]	New Orleans, LA/1997	Old housing, soil lead concentration/Association found between high soil lead areas and neighborhoods where children with EBLLs reside.
Choropleth mapping, overlay analysis, kriging, spatial autocorrelation/[21]	Syracuse, NY/1998	Old housing, race, population density, house value, rent/BLLs are correlated with percentage of children at risk, population density, mean housing value, and percentage of the African American population.
Overlay analysis, choropleth mapping/ [22]	Mexico/2002	Proximity to a point-source of lead exposure/There is a significant association between children with EBLLs and their distance to a point-source of lead exposure.
Address geocoding, overlay analysis/[23]	North Carolina/2007	Old housing, race, income, seasonality, water system/There is a correlation between water treatment systems and lead exposure among children.
Overlay analysis and kriging/[24]	New Orleans, LA/2011	Proximity to old and heavily used roads/Lead additives in gasoline had more impact on childhood lead exposure than the dust from leaded paint.

TABLE 1: *Cont.*

GIS Analysis/Citation	Region/Date	Common Risk Factors/Major Findings
Overlay analysis, buffer analysis, spatial masking/[25]	North Carolina/2011	Proximity to local airports/Significant positive association found between BLLs and the distances to the airport locations. Seasonality, age of housing, median household income and minority neighborhoods are also associated with BLLs.
Overlay analysis and Kriging/[26]	New Orleans, LA/2013	Soil lead concentrations in the old city core/A statistically significant relationship found between BLLs and soil lead level-proximity to old city cores.
Spatial analysis of genetic variation		
Choropleth mapping, overlay analysis/[27]	Durham, NC/2005	Race, and genetic vulnerability.
Political ecology		
Moran's I, LISA, and spatial autocorrelation/ [28]	North Carolina/2008	Old housing, poverty, tenant farming associated with the production of tobacco, rural African American population distribution.

Reviewed articles were summarized and grouped into five categories: screening methodology design, risk modeling studies, environmental risk factors, spatial analysis of genetic variation, and political ecology. Table 1 presents these studies under each category with GIS methods applied, study region, and common risk factors or major findings (Table 1). The first three categories focus on children's environment. The fourth category, spatial analysis of genetic variation, focuses on individual's traits. The last category, political ecology, focuses more on the long term socio-economic process of childhood lead poisoning. Some articles could fall into more than one category. We included articles into the categories where they mostly fit.

8.3 RESULTS AND DISCUSSION

All of the reviewed articles obtained their lead toxicity data from health departments. In these studies, blood lead screening data was collected by clinics or health workers without GIS. Data collection methods may vary among states.

8.3.1 SCREENING ACTIVITIES

Studies on childhood lead poisoning surveillance that used GIS include Lutz et al. in 1998, Reissman et al. in 2001, Roberts et al. in 2003, and Vaidyanathan et al. in 2009 [6,7,8,9]. These studies followed CDC's guidance on targeted screening [67]. The guidance requires that children at ages of 1 and 2 or ages of 3 and 6 should be tested if they have not been tested before and fall in at least one of the following criteria: residing in a ZIP code in which ≥27% of housing was built before 1950; receiving public assistance from programs such as Medicaid or the Special Supplemental Nutrition Program for Women, Infants and Children (WIC); and whose parents or guardians answer "yes" or "don't know" to at least one of the questions in basic personal-risk questionnaire.

The questions included in the questionnaire are: "Does your child live in or regularly visit a house that was built before 1950?"; "Does your child live or regularly visit a house built before 1978 with recent or ongoing renovations or remodeling within the last six months?"; and "Does your child have siblings or playmate who has or did have lead poisoning?" Some states had additional questions added to the CDC questionnaire. Lutz et al. [6] defined the "at-risk" population based on the questionnaire criteria. The study identified old housing and proximity to old roads as most common risk factors among those screened children. The authors produced three maps using the questionnaire data and census demographics. One of the maps shows the percentage of positive screenings for each census tract and another one displays "at-risk" and not "at-risk" screenings overlaid with the percentage of houses built before 1950. The third map plots EBLL children with the percentage of houses built before 1950. Although the study mapped the exact location of children, the state of Tennessee and some other states recently banned the disclosure of exact locations of the subjects in compliance with the HIPPA guidelines [68,69]. Lutz et al. found that the screening data thoroughly represents the targeted population in Knoxville, TN.

Reissman et al. [7] used GIS to assist the health department's decision making on screening activities in Louisville, Kentucky. The study attempts to (1) assess the efficacy of Jefferson County CLPP in surveying "at-risk" children and (2) determine the capability of GIS to find neighborhoods or housing units that pose risks to children. The first part of the study focuses on the childhood lead poisoning problem at the neighborhood level whereas the latter part examines the problem at the household level. Different from the Lutz et al. study [6], Reissman et al. considered the "at-risk" population as children between 6 and 35 months of age who reside in a home built before 1950 or live in a target zone where more than 27% of houses were built before 1950. The authors compared the percentage of screened children with corresponding target zones by both census tracts and ZIP codes. The study found that the percentage of children with EBLLs is strongly associated with old housing. The study also showed that the significant numbers of children who live in at risk areas were not being tested throughout the county. The second part of the study mapped the children who are younger than 7 years old with confirmed BLL ≥ 20 µg/dL and the houses where more than one child resides with confirmed BLL ≥ 20 µg/dL.

Roberts et al. [8] conducted a study over targeted lead-screening development using GIS in Charleston County, South Carolina. The authors obtained pediatric blood tests between 1991 and 1998 from Charleston County Lead Poisoning Prevention Program. Construction year of the houses was extracted from The Charleston County Tax Assessor. The authors first geocoded the children BLLs and then the buildings in the tax assessor by using Matchmaker/2000 address geocoding software. After the removal of duplicate building addresses from the tax assessor, the authors merged the two geocoded data sets: children BLLs and tax assessor buildings in Charleston County. Apart from Lutz et al. [6] and Reissman et al. [7], the authors categorized the housing variable in three categories; pre-1950, 1950–1977, and post-1977 in order to be consistent with the CDC's recommendations. Lead poisoning prevalence ratios in these time frames were compared. The study also displayed the actual locations of the children who have elevated blood lead levels (10 µg/dL and above). The study found that the children who live in a housing unit built before 1950 are four times more likely to have EBLLs than the children who live in a housing unit built after 1950. The study also found that there is no statistically significant difference between the children who live in a housing unit built between 1950 and 1977, and those who live in a housing unit built after 1977. In terms of screening activities, the study found that some areas with high number of pre-1950 housing were not screened at all.

Vaidyanathan et al. [9] developed a methodology to assess neighborhood risk factors for lead poisoning problems in Atlanta, Georgia in 2009. Unlike the studies referred in this section above, this study primarily used the Special Supplemental Nutrition Program for Women, Infants and Children (WIC) enrollments to identify "at-risk" populations. The authors used BLL data of children younger than 3 years of age when their blood was drawn in 2005. Three datasets were used in the study; pediatric blood tests by The Georgia Childhood Lead Poisoning Prevention Program (GA-CLPPP), the land parcel dataset for 1999 by the Center for GIS at the Georgia Institute of Technology in Atlanta, and census block group-level data from the 2000 US Census dataset. Since the boundary of block groups and neighborhoods did not coincide, the study followed a GIS methodology to transfer the age demographics from block groups to the neighborhood level in order to integrate residential land parcel data and blood lead tests with the demographics at the neighborhood level. The study indicated that only 11.9% of children aged ≤36 months from the city of Atlanta were tested for lead poisoning despite the risk of high lead exposure. The authors created a lead exposure index for the neighborhoods based on housing age and poverty. The poverty measure was calculated based on the number of children who were enrolled to the WIC. Housing age risk levels were composed of pre-1950 and pre-1978. The study reveals that 90% of residential units in Atlanta were built before 1978. These housing units might be an important source of lead exposure since most studies in the literature established a relationship between old housing and lead exposure through leaded paint. The study found that some neighborhoods are having as low as 8% of testing in children for lead poisoning whereas more than 78% of the children lived in housing units built before 1950. Excluding the Lutz et al. study, all of the studies in this section demonstrate that corresponding health departments failed to account for "at-risk" populations. The studies also demonstrate that GIS could be an effective tool to target "at-risk" neighborhoods by health departments.

8.3.2 RISK MODELING

This section refers to nine articles on risk model development for childhood lead poisoning [10,11,12,13,14,15,16,17,18]. Sargent et al. [10] conducted a census tract analysis over childhood lead exposure in Rhode Island. The study used 17,956 BLL screening records from the children who were aged 0 to 59 months and screened between 1992 and 1993. Because of the small area problem, the authors excluded two of the census tracts where there were very few screening samples. The study used the percentage of children with BLL $\geq$10 µg/dL as the dependent variable. The population of children for the census tracts was assigned based on census estimates. The study's final model includes five independent variables which explained 83% of the variance in lead exposure. According to the final model, percentages of screened children, households with public assistance income, houses built before 1950, vacant houses, and recent immigrants are positively associated with the outcome measure. Percentages of houses built before 1950 and vacant houses are significantly associated with the dependent variable. The source of lead exposure in immigrant children was unknown due to the possibility that they could be exposed to lead in their home countries. The study also found that there is no association between the percentage of African American population and high lead exposure in Rhode Island.

Miranda et al. [11] used a tax level address geocoding procedure to show high risk areas for North Carolina Childhood Lead Poisoning Prevention Program. The study covers the following North Carolina counties: Buncombe, Durham, Edgecombe, New Hanover, Orange, and Wilson. The authors first geocoded the screened children at the tax parcel unit in order to detect the age of housing from tax assessors datasets. Overall geocoding match rates vary from 47.2% to 72.1% for the six counties in North Carolina. Using this geocoded dataset, the authors employed analysis of variance (ANOVA) and multivariate analysis to find out whether the independent variables (age of the building, median income, and race) are statistically associated with the BLLs. Miranda et al. also prioritized the Durham, NC region in four risk areas: (1) predicted parcels which are most likely to contain leaded paint; (2) predicted parcels which are less likely to contain leaded paint; (3) predicted parcels which are lesser likely to contain leaded paint; and (4) predicted parcels which are least likely to contain leaded paint. Unlike the Sargent et al. study, Miranda et al. found that the dependent variable is correlated with the percentage of the African American population as well as median income and construction year of housings. One major shortcoming of the model is missing data since address geocoding rates may be under 50%. This study was later updated by Kim et al. [16] in 2008. The authors investigated how much the additional data from more intensive geocoding processes improved performance of childhood lead exposure risk models in identifying areas of elevated lead exposure. They used a comprehensive three-level stepwise address geocoding process. Similar to the studies by Miranda et al. [11], Griffith et al. [14] and Kim et al. also deployed an address geocoding based on the cadastral parcel reference system. Also similar to the Miranda et al. study [11], the geocoding success rate was lower because 31.2% of the addresses were not geocoded. The results in this study support the findings of the Miranda et al. [11] study and also find support for the following independent variables: percentage of Hispanic population, percentage of households with public assistance, and seasonality are also strongly associated with BLLs in the studied population.

Kriger et al. [12] examined temporal and spatial scale effects and the choice of geographical unit (i.e., census block group, census tract, and ZIP code) to monitor social inequalities in childhood lead poisoning. The authors used blood lead level screenings of children who live in Rhode Island. The screening period was between 1994 and 1996. Different from Miranda et al. [11], Kriger et al. used a street reference system (known as Topologically Integrated Geographic Encoding and Referencing (TIGER) dataset) for their address geocoding process. Street reference systems generally produce higher geocoding success rates compared to cadastral parcel reference systems. For instance, the Kriger et al. study produced more than 90% of geocoding success in all geographic units, census block groups, census tracts, and ZIP codes. However, one potential weakness of the method is that street geocoding results may be distant from the actual location of houses since the method uses a linear interpolation on street segments in the reference file. The authors found that the choice of measure and the level of geography matter. Census tract and census block group socioeconomic measures detected stronger socioeconomic gradients than the zip code units. The results indicate that BLLs are strongly associated with poverty but not education level, occupation, and wealth. A similar sensitivity analysis was conducted by Kaplowitz et al. [17] in 2010. Kaplowitz et al. assessed predictive validity of different geographic units for their risk assessment. According to their study, census block groups explain more variance in BLL than high and low risk ZIP codes. Their study confirmed that children's BLL is more closely associated with characteristics of their immediate environment than with characteristics of a larger area such as a census tract or ZIP code.

Haley and Talbot [13] presented a spatial analysis of BLLs in New York for the children born between 1994 and 1997. The study used the highest test result when there are multiple screens for a child. The authors obtained the birth records from the NYSDOH Bureau of Vital Statistics for the years between 1994 and 1997. Since the BLL records contain ZIP codes for the children, the authors used ZIP codes as the geographic units for spatial analysis. Apart from the other studies mentioned in this section, address geocoding was employed at ZIP level. Based on previous studies in the literature, Haley and Talbot selected the following socioeconomic variables: the percentage of houses built before 1940 and 1950, the percentage of adults ≥25 years of age who did not receive a high school diploma, the percentage of children living below the poverty level, the percentage of vacant housing units, the percentage of the population that rents a home, the percentage of the population screened in summer (July–September), population density, and the percentage of African-American births. The authors also used GIS to distribute the socio-economic data proportionally to the ZIP codes and to find the centroid locations of census blocks. In order to deal with missing data in the lead database, the authors used the mother's race from birth certificates and estimated the proportion of African American children for each ZIP code area. Unlike Sargent et al. [10], this study used a different methodology to deal with the small area problem. Using GIS, the authors merged the ZIP code areas when they have less than 100 screened children. Percentage of children with EBLLs in each ZIP code was defined as the dependent variable in the statistical analysis. The authors ran a multiple linear regression analysis to identify the relationship between the BLLs and the explanatory variables. They also analyzed the residuals' spatial autocorrelation in the model using

SpaceStat software and developed a simultaneous autoregressive model (SAR). Their regression analysis indicates that the age of housing, education level, and percentage of African American population variables are significant predictors of BLLs.

Griffith et al. [14] conducted an address geocoding study in 2007. The authors used BLLs data of children in Syracuse, NY between 1992 and 1996. The study compares two different address geocoding methods to find the impact of positional accuracy on spatial regression analysis of children's BLLs. These geocoding methods are based on street or polygon reference systems. The Haley and Talbot referred above used ZIP code boundaries as the polygon reference system. Griffith et al., on the other hand, used cadastral parcels as the reference files. Geocoding success rate is generally much higher in geocoding process with street reference files than ones with cadastral parcel reference files. However, cadastral parcel reference files provide more precise geocoding results and the construction year of housing units. The authors compared cadastral and TIGER based geocoded addresses in three sections including, census tract, census block group, and census blocks of 1990 and 2000 census demographics. The study shows that there is a noticeable but not considerably high positional error difference in their spatial statistical analyses using the two methods. The regression analysis in the study was employed in two different BLL thresholds, 5 and 10 μg/dL. Regardless of the threshold level, the results indicate that EBLLs are significantly associated with the percentage of the African American population and average house value in the census block and census block group analyses.

Using descriptive discriminant and odds ratio analyses, Oyana et al. [15,18] created a profile of high-risk areas based on housing age, the socioeconomic status, and ethnicity of the population in Chicago. The purpose of the study is to identify the health disparity among children who have different racial make-up. The study also assesses the spatiotemporal dynamics of the disease and identifies the socio-economic and racial composition of high-risk communities in Chicago. In addition, two different types of blood test methods (capillary and venous) were compared to one another for the BLL over 10 μg/dL. Oyana et al. uses a GIS scripting tool to deduplicate pediatric blood data. This study also differs from others by producing a kriging map for the area. The kriging map of Chicago shows that Westside area has the highest risk of EBLLs in the city. The authors also used TerraSeer's Space-Time Intelligence Systems (STIS) to explore the krigged prevalence rates in order to analyze spatial patterns [70]. Moran's I [71] and LISA statistics [72] were used with spatial autocorrelation to show the spatial patterns and health disparities in childhood lead toxicity in Chicago. The variations in raw prevalence rates for BLLs were high. However, kriging reduced the variations dramatically. The authors concluded that the dependent variable is significantly associated with housing age, income, and minority populations.

8.3.3 ENVIRONMENTAL RISK FACTORS

This section discusses eight studies that address environmental risk factors [19,20,21,22,23,24,25,26]. Guthe et al. [19] conducted one of the earliest GIS studies on childhood lead poisoning in 1992. The authors studied New Jersey municipalities of Newark, East Orange, and Irvington. The study mapped blood screening records overlaid with census tracts in the municipalities. Children blood samples were from the years 1983 to 1990. Unlike all the relevant studies reviewed in this article, the study used a 15 µg/dL threshold level, which was the BLL threshold level at the time. This study used street level address geocoding. Guthe et al. used command line address matching software, which is one of the oldest address geocoding engines. In terms of environmental factors, Mielke et al. [20] studied the associations between childhood BLLs and soil lead in Louisiana. The study used three data sets: soil lead data, age of housing data, and children blood lead data for urban New Orleans and rural Lafourche Parish in Louisiana. The study focused on soil contamination and leaded paint sources of lead toxicity problems. The percentage of housing built before 1940 was considered an indicator of leaded paint. Using x and y coordinates of census tract centroids, the authors plotted the three data sets within a three dimensional spatial model. The study showed that there is a relationship between low BLLs and new housing neighborhoods, and old housing neighborhoods were split evenly between old and new housing. There is also an association between high soil lead areas and neighborhoods where children with EBLLs reside. The study suggests that inner-city children should be the focus area to eliminate lead toxicity problems in the population.

Griffith et al. [21] employed several GIS tools that include geocoding, buffer analysis, and interpolation techniques such as kriging to depict the lead poisoning problem in Syracuse, NY. This study shows the geographic distribution of lead toxicity in Syracuse, NY in three aggregated levels: census block, census block group, and census tract. Linear regression with spatial autocorrelation is used as a statistical method for the three aggregated levels. The study shows that there is a major difference between urban and rural exposure, which is consistent with the results from Laidlaw et al., and Mielke et al. [56,58,59]. It however finds no statistically significant relationship between historically heavily traveled streets and lead exposure. Lead poisoning is detectable regardless of the level of geographic resolution. Griffith et al. also showed that BLLs are correlated with percentage of children at risk, population density, mean housing value, and percentage of the African American population.

Gonzalez et al. [22] investigated the possible impact of point sources of lead exposure relative to other types of lead exposure sources. The study was conducted in Tijuana, Mexico with, Hispanic children aged between 1.5 and 6.9 years. In order to deal with the confounding variable of cultural habits, the study used BLLs where the subjects reported that they did not use lead-glazed ceramics for cooking or food storage purposes. The study was composed of 76 samples from 14 sites. Gonzalez et al. mapped the distribution of these 76 point sources as well as five point sources containing 19 soil samples with the values ranging from 100 to 7870 µg/g soil lead. They compared the children BLLs with Bocco and Sachez [73] study's prediction model which was based on fixed industrial lead point sources. Similar to the Bocco and Sachez study, the authors assigned Tijuana census tracts the labels of "high", "medium", "low", and "N/A" risk levels based on proximity to the lead point sources. The authors also mapped these risk levels of census tracts and children cases with elevated blood lead levels (≥ 10 µg/dL) where the subjects reported non-use of lead-glazed ceramics.

In 2007, Miranda et al. [23] explored the potential effect of the use of chloramines in water treatment systems over childhood lead exposure in Wayne County, North Carolina. The authors examined the relationship between these potential effects and the age of housing in order to help guide policy practices in North Carolina. The authors used the datasets of children BLLs, tax parcels, census data, and water treatment system boundaries. Children BLLs were geocoded based on tax parcels with a 72.4% geocoding success rate from the surveillance data between 1999 and 2003. The study used multivariate regression to analyze the data and concluded that the use of chloramines in the water treatment systems might inadvertently increase lead exposure among children.

Another environmental study by Miranda et al. [25] conducted in 2011 to investigate the relationship between avgas lead exposure and children BLLs. The authors selected 66 airports in 6 counties of North Carolina based on the availability of tax assessor data, the volume of air traffic, and the number of screened children for lead toxicity. The study used the airports' estimated annual lead emissions which were obtained from the U.S. EPA Office of Transportation and Air Quality. The children BLL data composed of the blood tests conducted between 1995 and 2003 for the children between the ages of 9 months and 7 years. The authors determined the airport boundaries using tax parcel data. The authors created buffer zones surrounding each airport selected in the study. The buffers were created based on the distances of 500 m, 1000 m, 1500 m, and 2000 m from the polygon edges of the airports. Unlike most of the studies discussed in this review article, Miranda et al. used GIS to show children locations in a jittered representation even though they run the statistical model based on actual point locations. Using the geocoded locations, Miranda et al. was able to join children locations and buffer zones, which were created from the airport boundaries. The authors assigned dummy variables to children locations based on the boundaries mentioned above and seasons for the screening time. The model includes the age of housing, screening season, and demographic variables. The authors also used inverse population weights to eliminate the possible bias caused by high numbers of screening cases on parcels. The study found a significant positive association between logged BLLs and the distances to the airport locations. It further shows that seasonality is an important factor in estimating BLLs. In fall, spring, and summer seasons, children were found having higher BLLs on average compared to winter season screenings. Age of housing was negatively associated with BLLs while the median household income and minority neighborhoods had positive associations with BLLs.

Mielke et al. [24] conducted a comparative analysis of lead poisoning problems by assessing the pre-Katrina blood and soil lead concentrations around public and private properties in New Orleans. Soil lead data was composed of 587 soil samples (224 samples from public properties, and 363 samples from residential private properties) and 55,551 BLL screening records for the years between 2000 and 2005. The study shows significant differences among the blood lead prevalence between the inner city (CJ Peete) and outlying areas (Florida) of New Orleans. The study also found no statistically significant other differences between inner and outer cities. The authors found that, among the screens in public properties, differences between inner and outer cities in lead toxicity prevalence are a better proxy than age of construction. The study noted that lead additives in gasoline had more impact on childhood lead exposure than the dust from leaded paint. In terms of lead dust from vehicles, the largest amount of lead was deposited on soil in the inner-cities whereas outer-cities were not experiencing a large amount of lead deposit from the exhaust due to a lighter traffic volume. Consequently, the study indicated that lead toxicity originated from soil contamination could help explain lead toxicity in children.

In 2013, Mielke et al. [26] analyzed the association between children blood lead levels and soil lead concentrations in relation to before and after hurricane Katrina in New Orleans. In the study, pre-Katrina was from 2000 to 2005 and post-Katrina was from 2006 to 2008. Children's blood samples (55,551 records in pre-Katrina and 7384 records in post-Katrina period) were geocoded at the 1990 census tract level. Soil lead data was composed of 5467 soil samples. Soil samples were categorized by their one meter proximity to "busy streets", "residential streets", "house sides", and "open spaces". Census tract medians of soil lead concentration data were used to produce kriging maps of soil lead concentration for both pre- and post-Katrina periods. Census tracts were also categorized as low and high in lead concentration groups based on 100 mg/kg threshold (≥ 100 mg/kg and <100 mg/kg). Non-parametric statistics were used because of positive skewness in the soil lead data. Multi-purpose permutation procedure showed that there was a significant difference between low and high lead tracts. This confirms the significance of 100 mg/kg as a threshold for lead concentration in soil for New Orleans. Census tract soil lead concentration medians showed that busy streets had the highest median by location. This could be related to historical lead deposits from car exhausts. Kriging maps showed that there was no major change in the lead concentration level in soil for pre- and post-Katrina periods. Unlike Griffith et al. [21], this study suggests that there is a statistically significant relationship between BLLs and soil lead level-proximity to old city cores.

8.3.4 GENETIC VARIATION

One of the reviewed studies focused on the genetic variation of childhood lead poisoning problems [27]. Since other studies found a significant relationship between childhood lead poisoning and African American populations, the authors focused on genetic variation of the problem. The study used previously developed data of children BLLs by Miranda et al. [11], which geocoded children cases at the tax parcel level in order to get the construction year of house units from tax assessor data. The study also considers the occupancy status, which was also gathered from tax parcels. The authors note that the spatial autocorrelation problems were minimized by assigning individual year of construction from tax parcels. The ANOVA comparison of models with and without spatial autocorrelation also corroborated the non-existence of spatial autocorrelation. Since some of the information pertaining to construction years is missing in the tax parcel dataset, some cases lacked this information. In those cases, the study assigned the construction year from the nearby parcels. Some studies in the literature indicate that the relationship between high BLLs and African American populations might be because of low calcium intake in the population. According to this study, however, the relationship between high BLLs and African American populations might be more related to genetic polymorphisms.

8.3.5 POLITICAL ECOLOGY

Hanchette's study focused on the political ecology aspect of childhood lead toxicity. The author used Moran's I [71] and LISA statistics [72] to investigate the spatial distribution of lead poisoning prevalence at the county level in North Carolina. They use 10-year children BLL data from 1995 to 2004. In the study, the data findings show that there is a significant cluster of high BLL rates in eastern North Carolina. The author indicated that these clusters of high rates show persistent health disparities in the region. Hanchette claims that the health disparities in eastern North Carolina results from large scale socio-economic and cultural processes rather than neighborhood characteristics such as poverty and old housing. The study found that the Appalachia (western North Carolina) region displayed low rates of lead poisoning even though the region had high poverty rates. Another major finding is that high rates of lead poisoning clusters correspond with African American populations only in eastern North Carolina. Unlike this region, southern North Carolina does not have high rates of lead poisoning despite high concentration of African American populations. The author suggests that the convergence of poverty, older housing, and the large rural African American population can be explained by the long history of tenant farming. According to Hanchette, this transition from an agricultural state to a mixed economy led to changes in socio-economic characteristics of the eastern region of North Carolina.

8.4 CONCLUSIONS

This article reviewed twenty-three GIS-based studies examining spatial modeling of childhood lead poisoning and risk factors that were published after 1991, the year the CDC's threshold updated to10 µg/dL. GIS use in lead studies revealed greater detail about the magnitude of lead poisoning within populations. Reviewed articles indicate that surveillance and screening practices have extended considerable amount of importance in targeting "at-risk" populations. However, the literature shows that some health departments failed to account for "at-risk" populations [7,8,9]. This issue can be resolved through the implementation of GIS in health departments.

Risk factors for childhood lead poisoning (age of housing, urban/rural status, race/ethnicity, socioeconomic status, population density, renter/owner occupancy, housing value, and nutritional status) have been thoroughly parsed out in childhood lead poisoning research. Unfortunately, address geocoding methods, the parameters used, and the uncertainties they presented were not included in a similar level of detail in the research. Most of the reviewed studies did not provide the input parameters such as the reference system and the match rate. Since these parameters have a direct impact on results of the spatial analyses, this makes it difficult to conduct legitimate comparisons among the various articles.

Even though to date no safe blood lead thresholds for the adverse effects of lead on children have been identified [31], data related to children with very low BLLs has consistently been overlooked. Address information of children with BLLs ranging from 0–3 µg/dL may not be reported since screening efforts have primarily focused on children with high BLLs [74]. This non-random missing data can cause misinterpretation of the spatial distribution of lead poisoning. In order to improve the quality of geocoding, the addresses need to be confirmed in the data collection phase of a GIS environment. Such GIS-integrated screening could eliminate spatial bias due to disparities in reporting. Future studies are needed to fill this gap and attempt to improve the use of address geocoding in BLL data collection.

Future lead poisoning studies should also be concerned with data aggregation and the choice of geographical analysis. Data aggregation is done for two reasons: to link socio-economic and environmental measures to lead data and to ensure data confidentiality. In the former case, geocoded addresses may fall far away from their actual locations resulting in boundary problems during data aggregation to census block groups, census tracts, or ZIP code areas. Very few studies examined these aggregation problems and spatial scale effects to monitor risk factors [14]. Studies show that finer geographic units such as census block group levels explain lead poisoning problems better, and hence some high levels of data aggregation (such as ZIP codes or census tracts) may not explain the distribution in the population [12,17]. Moreover, longitudinal lead studies are subject to possible errors as a result of change in census boundaries over time. In the latter case, very few studies examined the use of GIS and developed techniques to preserve confidentiality during the process of dissemination of screened children data and the resultant high risk areas [25].

Environmental studies on lead paint usage before 1978 have shown a link between house age and elevated BLLs. Soil studies can also reveal sources of lead toxicity. Several studies have shown that the distribution of lead toxicity among young children can be explained by proximity to high volume traffic areas. The relationship of vehicular lead deposits and children with elevated BLLs is contentious. Griffith et al. [21] found no relationship between childhood lead toxicity and their proximity to heavily traveled roads. Contrary to Griffith's findings, Mielke et al. [20,24,26] found that childhood lead poisoning was related to residing in inner-city areas where the traffic flow was historically larger. Miranda et al. [25] also found a correlation between the proximity of airports and BLLs among children. None of the reviewed studies accounted for housing abatement efforts in their models. Future studies focusing on environmental lead sources need to factor in abatement efforts that may have taken place. By factoring in housing abatement efforts we can eliminate erroneous data and misinterpretations.

The environmental studies in this review also indicate a correlation between BLLs and African American populations. However, very few studies investigated the individual characteristics of children [27]. The history of socioeconomic and cultural processes could also be important factors to identify risk areas [28]. More GIS-based studies need to be conducted to investigate these factors. All of the articles reviewed in this paper show the development of an increasing awareness of the intricacies of lead poisoning and its effects on children and their neighborhoods.

REFERENCES

1. Using GIS to Assess and Direct Childhood Lead Poisoning Prevention: Guidance for State and Local Childhood Lead Poisoning Prevention Programs. Dec, 2004. [(accessed on 21 February 2014)]. Available online: http://www.cdc.gov/nceh/lead/publications/UsingGIS.pdf.
2. Healthy People 2020 Objectives. [(accessed on 24 February 2014)]. Available online: http://healthypeople.gov/2020/topicsobjectives2020/overview.aspx?topicid=12.
3. Yasnoff W.A., Sondik E.J. Geographic Information Systems (GIS) in public health practice in the new millennium. J. Public Health Manag. Pract. JPHMP. 1999;5:ix–xii. doi: 10.1097/00124784-199907000-00001.
4. Health N.C. For E. CDC—Lead—Funding. [(accessed on 15 May 2014)]. Available online: http://www.cdc.gov/nceh/lead/funding.htm.
5. CDC Response to Advisory Committee on Childhood Lead Poisoning Prevention Recommendations in "Low Level Lead Exposure Harms Children: A Renewed Call

of Primary Prevention" [(accessed on 21 February 2014)]. Available online: http://www.cdc.gov/nceh/lead/acclpp/cdc_response_lead_exposure_recs.pdf.

6. Lutz J., Jorgensen D., Hall S., Julian J. Get the lead out: A regional approach to healthcare and beyond. Geo. Info. Syst. 1998;8:26–30.

7. Reissman D.B., Staley F., Curtis G.B., Kaufmann R.B. Use of geographic information system technology to aid Health Department decision making about childhood lead poisoning prevention activities. Environ. Health Perspect. 2001;109:89–94. doi: 10.1289/ehp.0110989.

8. Roberts J.R., Hulsey T.C., Curtis G.B., Reigart J.R. Using geographic information systems to assess risk for elevated blood lead levels in children. Public Health Rep. 2003;118:221–229. doi: 10.1016/S0033-3549(04)50243-1.

9. Vaidyanathan A., Staley F., Shire J., Muthukumar S., Kennedy C., Meyer P.A., Brown M.J. Screening for lead poisoning: A geospatial approach to determine testing of children in at-risk neighborhoods. J. Pediatr. 2009;154:409–414. doi: 10.1016/j.jpeds.2008.09.027.

10. Sargent J.D., Bailey A., Simon P., Blake M., Dalton M.A. Census tract analysis of lead exposure in Rhode island children. Environ. Res. 1997;74:159–168. doi: 10.1006/enrs.1997.3755.

11. Miranda M.L., Dolinoy D.C., Overstreet M.A. Mapping for prevention: GIS models for directing childhood lead poisoning prevention programs. Environ. Health Perspect. 2002;110:947–953.

12. Krieger N., Chen J., Waterman P., Soobader M., Subramanian S., Carson R. Choosing area based socioeconomic measures to monitor social inequalities in low birth weight and childhood lead poisoning: The Public Health Disparities Geocoding Project (US) J. Epidemiol. Community Health. 2003;57:186–199. doi: 10.1136/jech.57.3.186.

13. Haley V.B., Talbot T.O. Geographic analysis of blood lead levels in New York state children born 1994–1997. Environ. Health Perspect. 2004;112:1577–1582. doi: 10.1289/ehp.7053.

14. Griffith D.A., Millones M., Vincent M., Johnson D.L., Hunt A. Impacts of positional error on spatial regression analysis: A case study of address locations in Syracuse, New York. Trans. GIS. 2007;11:655–679. doi: 10.1111/j.1467-9671.2007.01067.x.

15. Oyana T.J., Margai F.M. Geographic analysis of health risks of pediatric lead exposure: A golden opportunity to promote healthy neighborhoods. Arch. Environ. Occup. Health. 2007;62:93–104. doi: 10.3200/AEOH.62.2.93-104.

16. Kim D., Galeano M.A.O., Hull A., Miranda M.L. A framework for widespread replication of a highly spatially resolved childhood lead exposure risk model. Environ. Health Perspect. 2008;116:1735–1739. doi: 10.1289/ehp.11540.

17. Kaplowitz S.A., Perlstadt H., Post L.A. Comparing lead poisoning risk assessment methods: Census block group characteristics vs. zip codes as predictors. Public Health Rep. 2010;125:234–245.

18. Oyana T.J., Margai F.M. Spatial patterns and health disparities in pediatric lead exposure in chicago: Characteristics and profiles of high-risk neighborhoods. Prof. Geogr. 2010;62:46–65. doi: 10.1080/00330120903375894.

19. Guthe W.G., Tucker R.K., Murphy E.A., England R., Stevenson E., Luckhardt J.C. Reassessment of lead exposure in New Jersey using GIS technology. Environ. Res. 1992;59:318–325. doi: 10.1016/S0013-9351(05)80038-6.

20. Mielke H.W., Dugas D., Mielke P.W., Jr., Smith K.S., Smith S.L., Gonzales C.R. Associations between Soil lead and childhood blood lead in urban New Orleans and Rural Lafourche Parish of Louisiana. Environ. Health Perspect. 1997;105:950–954. doi: 10.1289/ehp.97105950.

21. Griffith D.A., Doyle P.G., Wheeler D.C., Johnson D.L. A tale of two swaths: Urban childhood blood-lead levels across Syracuse, New York. Ann. Assoc. Am. Geogr. 1998;88:640–665. doi: 10.1111/0004-5608.00116.

22. Gonzalez E.J., Pham P.G., Ericson J.E., Baker D.B. Tijuana childhood lead risk assessment revisited: Validating a GIS model with environmental data. Environ. Manag. 2002;29:559–565. doi: 10.1007/s00267-001-0007-1.

23. Miranda M.L., Kim D., Hull A.P., Paul C.J., Galeano M.A.O. Changes in blood lead levels associated with use of chloramines in water treatment systems. Environ. Health Perspect. 2007;115:221–225.

24. Mielke H.W., Gonzales C.R., Mielke P.W., Jr. The continuing impact of lead dust on children's blood lead: Comparison of public and private properties in New Orleans. Environ. Res. 2011;111:1164–1172. doi: 10.1016/j.envres.2011.06.010.

25. Miranda M.L., Anthopolos R., Hastings D. A geospatial analysis of the effects of aviation gasoline on childhood blood lead levels. Environ. Health Perspect. 2011;119:1513–1516. doi: 10.1289/ehp.1003231.

26. Mielke H.W., Gonzales C.R., Powell E.T., Mielke P.W. Environmental and health disparities in residential communities of New Orleans: The need for soil lead intervention to advance primary prevention. Environ. Int. 2013;51:73–81. doi: 10.1016/j.envint.2012.10.013.

27. Miranda M.L., Dolinoy D.C. Using GIS-based approaches to support research on neurotoxicants and other children's environmental health threats. Neurotoxicology. 2005;26:223–228. doi: 10.1016/j.neuro.2004.10.003.

28. Hanchette C.L. The political ecology of lead poisoning in eastern North Carolina. Health Place. 2008;14:209–216.

29. WHO Childhood Lead Poisoning. [(accessed on 16 May 2014)]. Available online: http://www.who.int/ceh/publications/childhoodpoisoning/en/

30. McDonald J.A., Potter N.U. Lead's legacy? Early and late mortality of 454 lead-poisoned children. Arch. Environ. Health. 1996;51:116–121. doi: 10.1080/00039896.1996.9936003.

31. Brown M.J., Margolis S. Lead in drinking water and human blood lead levels in the United States. MMWR. 2012;61:1–9.

32. Graff J.C., Murphy L., Ekvall S., Gagnon M. In-home toxic chemical exposures and children with intellectual and developmental disabilities. Pediatr. Nurs. 2006;32:596–603.

33. Reed A.J. Lead poisoning—Silent epidemic and social crime. Am. J. Nurs. 1972;72:2180–2184. doi: 10.2307/3422489.

34. Chisolm J.J., Mellits E.D., Keil J.E., Barrett M.B. Variations in hematologic responses to increased lead absorption in young children. Environ. Health Perspect. 1974;7:7–12. doi: 10.1289/ehp.7477.

35. Cooper M.H. Lead Poisoning. [(accessed on 20 February 2014)]. Available online: http://library.cqpress.com/cqresearcher/cqresrre1992061900.
36. Edwards M. Lead poisoning: A public health issue. Prim. Health Care. 2008;18:18.
37. Bochynska, Katarzyna Facts and Firsts of Lead. [(accessed on 16 May 2014)]. Available online: http://www.lead.org.au/fs/fst29.html.
38. United States Environmental Protection Agency Toxic Substances Control Act (TSCA) [(accessed on 13 June 2014)]. Available online: http://www.epa.gov/oecaagct/lsca.html#Lead-Based%20Paint.
39. United States Environmental Protection Agency Basic Information about Lead in Drinking Water. [(accessed on 13 June 2014)]. Available online: http://water.epa.gov/drink/contaminants/basicinformation/lead.cfm.
40. United States Environmental Protection Agency Basic Information. Fuels and Fuel Additives. [(accessed on 25 February 2014)]. Available online: http://www.epa.gov/otaq/fuels/basicinfo.htm.
41. Edwards M. Fetal death and reduced birth rates associated with exposure to lead-contaminated drinking water. Environ. Sci. Technol. 2013 doi: 10.1021/es4034952.
42. Trasande L., Liu Y. Reducing the Staggering costs of environmental disease in children, estimated at $76.6 billion in 2008. Health Aff. (Millwood) 2011;30:863–870. doi: 10.1377/hlthaff.2010.1239.
43. Preventing Lead Poisoning in Young Children: A Statement by the Centers for Disease Control. Mar, 1975. [(accessed on 21 February 2014)]. Available online: http://www.cdc.gov/nceh/lead/publications/plpyc1975.pdf.
44. Preventing Lead Poisoning in Young Children: A Statement by the Centers for Disease Control. Apr, 1978. [(accessed on 21 February 2014)]. Available online: http://www.cdc.gov/nceh/lead/publications/plpyc1978.pdf.
45. Preventing Lead Poisoning in Young Children: A Statement by the Centers for Disease Control. Jan, 1985. [(accessed on 21 February 2014)]. Available online: http://www.cdc.gov/nceh/lead/publications/plpyc1985.pdf.
46. Preventing Lead Poisoning in Young Children: A Statement by the Centers for Disease Control. Oct, 1991. [(accessed on 21 February 2014)]. Available online: http://www.cdc.gov/nceh/lead/Publications/books/plpyc/contents.htm.
47. Preventing Lead Poisoning in Young Children: A Statement by the Centers for Disease Control and Prevention. Aug, 2005. [(accessed on 21 February 2014)]. Available online: http://www.cdc.gov/nceh/lead/publications/prevleadpoisoning.pdf.
48. Wartenberg D. Screening for lead exposure using a geographic information system. Environ. Res. 1992;59:310–317. doi: 10.1016/S0013-9351(05)80037-4.
49. Litaker D., Kippes C.M., Gallagher T.E., O'Connor M.E. Targeting lead screening: The Ohio Lead Risk Score. Pediatrics. 2000;106:e69:1–e69:8.
50. Joseph C.L.M., Havstad S., Ownby D.R., Peterson E.L., Maliarik M., McCabe M.J., Barone C., Johnson C.C. Blood lead level and risk of asthma. Environ. Health Perspect. 2005;113:900–904. doi: 10.1289/ehp.7453.
51. Laidlaw M.A.S., Mielke H.W., Filippelli G.M., Johnson D.L., Gonzales C.R. Seasonality and children's blood lead levels: Developing a predictive model using climatic variables and blood lead data from Indianapolis, Indiana, Syracuse, New York, and New Orleans, Louisiana (USA) Environ. Health Perspect. 2005;113:793–800. doi: 10.1289/ehp.7759.

52. Lo Y.-C., Dooyema C.A., Neri A., Durant J., Jefferies T., Medina-Marino A., de Ravello L., Thoroughman D., Davis L., Dankoli R.S., et al. Childhood lead poisoning associated with gold ore processing: A village-level investigation—Zamfara State, Nigeria, October–November 2010. Environ. Health Perspect. 2012;120:1450–1455. doi: 10.1289/ehp.1104793.

53. Graber L.K., Asher D., Anandaraja N., Bopp R.F., Merrill K., Cullen M.R., Luboga S., Trasande L. Childhood lead exposure after the phaseout of leaded gasoline: An ecological study of school-age children in Kampala, Uganda. Environ. Health Perspect. 2010;118:884–889. doi: 10.1289/ehp.0901768.

54. Farfel M.R., Orlova A.O., Lees P.S.J., Rohde C., Ashley P.J., Chisolm J.J. A study of urban housing demolitions as sources of lead in ambient dust: Demolition practices and exterior dust fall. Environ. Health Perspect. 2003;111:1228–1234. doi: 10.1289/ehp.5861.

55. Navas-Acien A., Guallar E., Silbergeld E.K., Rothenberg S.J. Lead exposure and cardiovascular disease—A systematic review. Environ. Health Perspect. 2007;115:472–482.

56. Laidlaw M.A.S., Filippelli G.M. Resuspension of urban soils as a persistent source of lead poisoning in children: A review and new directions. Appl. Geochem. 2008;23:2021–2039. doi: 10.1016/j.apgeochem.2008.05.009.

57. Brown R.W., Longoria T. Multiple risk factors for lead poisoning in hispanic subpopulations: A review. J. Immigr. Minor. Health. 2010;12:715–725. doi: 10.1007/s10903-009-9245-8.

58. Mielke H.W., Laidlaw M.A.S., Gonzales C. Lead (Pb) legacy from vehicle traffic in eight California urbanized areas: Continuing influence of lead dust on children's health. Sci. Total Environ. 2010;408:3965–3975. doi: 10.1016/j.scitotenv.2010.05.017.

59. Mielke H.W., Laidlaw M.A.S., Gonzales C.R. Estimation of leaded (Pb) gasoline's continuing material and health impacts on 90 US urbanized areas. Environ. Int. 2011;37:248–257. doi: 10.1016/j.envint.2010.08.006.

60. Laidlaw M.A.S., Taylor M.P. Potential for childhood lead poisoning in the inner cities of Australia due to exposure to lead in soil dust. Environ. Pollut. 2011;159:1–9. doi: 10.1016/j.envpol.2010.08.020.

61. Filippelli G.M., Laidlaw M.A.S., Latimer J.C., Raftis R. Urban lead poisoning and medical geology: An unfinished story. GSA Today. 2005;15 doi: 10.1130/1052-5173.

62. Lejano R., Ericson J. Tragedy of the temporal commons: Soil-bound lead and the anachronicity of risk. J. Environ. Plan. Manag. 2005;48:301–320. doi: 10.1080/0964056042000338190.

63. Mielke H.W., Gonzales C., Powell E., Mielke P.W., Jr. Urban soil-lead (Pb) footprint: Retrospective comparison of public and private properties in New Orleans. Environ. Geochem. Health. 2008;30:231–242. doi: 10.1007/s10653-007-9111-3.

64. Díaz-Barriga F., Batres L., Calderón J., Lugo A., Galvao L., Lara I., Rizo P., Arroyave M.E., McConnell R. The El Paso smelter 20 years later: Residual impact on Mexican children. Environ. Res. 1997;74:11–16. doi: 10.1006/enrs.1997.3741.

65. Simpson E., Mull J.D., Longley E., East J. Pica during pregnancy in low-income women born in Mexico. West. J. Med. 2000;173:20–24. doi: 10.1136/ewjm.173.1.20.

66. Hamilton S., Rothenberg S.J., Khan F.A., Manalo M., Norris K.C. Neonatal lead poisoning from maternal pica behavior during pregnancy. J. Natl. Med. Assoc. 2001;93:317–319.

67. Centers for Disease Control and Prevention . Atlanta: 1997. [(accessed on 18 May 2014)]. Screening Young Children for Lead Poisoning: Guidance for State and Local Public Health Officials. Available online: http://www.cdc.gov/nceh/lead/publications/1997/pdf/chapter3.pdf.

68. Summary of the HIPAA Privacy Rule. [(accessed on 17 May 2014)]. Available online: http://www.hhs.gov/ocr/privacy/hipaa/understanding/summary/index.html.

69. Summary of the HIPAA Security Rule. [(accessed on 17 May 2014)]. Available online: http://www.hhs.gov/ocr/privacy/hipaa/understanding/srsummary.html.

70. Jacquez G.M. Space-time Intelligence System Software for the Analysis of Complex Systems. In: Fischer M.M., Getis A., editors. Handbook of Applied Spatial Analysis. Springer; Berlin/Heidelberg, Germany: 2010. pp. 113–124.

71. Moran P.A.P. Notes on continuous stochastic phenomena. Biometrika. 1950;37:17–23. doi: 10.1093/biomet/37.1-2.17.

72. Anselin L. Local indicators of spatial association—LISA. Geogr. Anal. 1995;27:93–115. doi: 10.1111/j.1538-4632.1995.tb00338.x.

73. Bocco G., Sánchez R. Identifying Potential impact of lead contamination using a geographic information system. Environ. Manag. 1997;21:133–138. doi: 10.1007/s002679900012.

74. Betsy Shockley. (Shelby County Health Department, Memphis, TN, USA. Childhood Lead Poisoning Prevention Program in Tennessee). Personal interview. Sep 27, 2013.

CHAPTER 9

LINKING SOURCE AND EFFECT: RESUSPENDED SOIL LEAD, AIR LEAD, AND CHILDREN'S BLOOD LEAD LEVELS IN DETROIT, MICHIGAN

SAMMY ZAHRAN, MARK A.S. LAIDLAW, SHAWN P. MCELMURRY, GABRIEL M. FILIPPELLI, AND MARK TAYLOR

9.1 INTRODUCTION

Lead (Pb) remains a serious threat to children's health and development—elevated levels of Pb in the blood are associated with impaired cognitive, motor, behavioral, and physical abilities.(1) Even lead-exposed children with blood lead levels (BLLs) below the World Health Organization (WHO) guideline of 10 µg/dL for their entire lifetime experience measurable loss in cognition.(2) In response to health risks associated with BLLs below 10 µg/dL,(3-5) the U.S. Centers for Disease Control and Prevention (CDC) lowered the blood Pb reference value to 5 µg/dL in May 2012.

Average BLLs in the U.S. (and globally) declined following the elimination of Pb from most product streams (e.g., gasoline, paint, water pipes, and solder used to seal canned goods). While airborne Pb used to be extremely high in cities, largely from the direct combustion of leaded gaso-

Reprinted with permission from Zahran S, Laidlaw MAS, McElmurry SP, Filippelli GM, and Taylor M. Linking Source and Effect: Resuspended Soil Lead, Air Lead, and Children's Blood Lead Levels in Detroit, Michigan. Environmental Science and Technology *47,6 (2013), DOI: 10.1021/es303854c. Copyright 2013, American Chemical Society.*

line and deposition of Pb oxides, much of the current airborne Pb is from these legacy sources. Contemporary air Pb is in the form of resuspended fine particulates.(6-10)

In this paper, we aim to explain the lingering sources of Pb in Detroit, Michigan by analytically reconciling a compelling empirical fact: average BLLs for children in the northern hemisphere peak in summer and autumn and retreat during winter and spring periods.(11) In Detroit, the BLLs of children follow this seasonal phenomenon (see Supporting Information (SI)).

As compared to the reference month of January, child BLLs are found to be between 11% and 14% higher in the months of July, August, and September (described in detail in Figure SI1 and Table SI1 of the SI). The seasonal behavior of child BLLs in Detroit is clear. Explaining this seasonal phenomenon is the aim of our paper, and it is our contention that any theory of contemporary Pb risk must logically account for this striking empirical observation.

Our intuition of what could plausibly account for the seasonality of child BLLs in Detroit is derived from a series of known facts. First, similar to many other postindustrial cities, elevated concentrations of environmental Pb are found throughout the Detroit metropolitan area,(12) with especially high concentrations of Pb in soils (400–800 mg/kg) located in the interior of the city(13) that correlate with spatial variation in children's BLLs. (14) Outdoor soil is a reservoir of legacy Pb from multiple anthropogenic sources and may explain why many household intervention efforts are unsuccessful.(15) Second, research shows that contemporary atmospheric concentrations of Pb spike during summer and autumn in many U.S. cities, including Washington, DC,(16, 17) Boston,(18) Milwaukee,(19) New York,(20) New Jersey,(21) and Chicago.(22) In fact, seasonal variations in child BLLs and atmospheric Pb are strikingly similar. Third, previous research has demonstrated a remarkable ability to predict child BLLs based on climate variables.(23) Taken together, these facts are suggestive of a soil → air dust → child pathway of contemporary Pb exposure, where Pb-contaminated urban soils are resuspended as dust subject to seasonal precipitation regimes, wind, humidity, and other meteorological factors, with air Pb dust inhaled and ingested by unsuspecting children.

To evaluate this hypothesized pathway, this study uses temporally resolved atmospheric soil and Pb data, and matched BLL data from the Detroit metropolitan area. Statistical and numerical modeling are used to determine correlation strengths across a range of environmental and human variables, and specifically to target the contributions of air Pb to child BLLs likely derived from soil resuspension as opposed to point source air Pb emissions.

9.2 MATERIALS AND METHODS

To address the soil → air dust → child pathway for Pb exposure, a number of data sources are examined: blood Pb data for 367 839 children from the Michigan Department of Community Health (MDCH); atmospheric soil and Pb data from the U.S. Environmental Protection Agency's (EPAs) Interagency Monitoring of Protected Visual Environments (IMPROVE) database; (24) local weather data from the National Weather Service, National Climatic Data Center; and point location information on Pb-emitting facilities from the EPA's Toxic Release Inventory.(25)

9.2.1 BLOOD PB DATA

Children's blood Pb data for the tricounty area encompassing the City of Detroit was obtained from the Michigan Department of Community Health (MDCH). The data set contains blood samples collected from January 2001 through December 2009. Blood Pb measurements are reported as integers in units of micrograms per deciliter of blood (μg/dL). MDCH data also contain information on the census tract residential location of each child, the month and year of sample collection, child age in years (0–10), child sex (male = 1, female = 0), and the blood draw type (1 = capillary, 0 = venous). As with previous research,(26) we analyze child BLLs as a continuous variable (μg/dL) and as dichotomous variable (≥ 5 μg/dL = 1, < 5 μg/dL = 0).

9.2.2 ATMOSPHERIC PB AND SOIL DATA

Atmospheric soil and Pb aerosol data were obtained from IMPROVE for the period of January 2001 to December 2009 (Station 261630001; additional stations presented in the SI). To derive atmospheric soil estimates, we use a mineral equation based on the elemental composition of soil.(27) Soil composition is derived by the quadratic sum of aluminum (Al), silica (Si), calcium (Ca), iron (Fe), and titanium (Ti) concentrations, assuming independence of measurement uncertainties as described by:

$$[d(soil)]^2 = [2.20 \times d(Al)]^2 + [2.49 \times d(Si)]^2 + [1.63 \times d(Ca)]^2 + [2.42 \times d(Fe)]^2 + [1.94 \times d(Ti)]^2 \tag{1}$$

Both atmospheric soil and Pb aerosol quantities are measured in units of $\mu g/m^3$. The quantity of soil derived using this equation is an estimate subject to spatial variability of soil composition and anthropogenic interferences.

9.2.3 LOCAL WEATHER DATA

Given that local weather conditions influence atmospheric concentrations of soil and Pb, we collected data describing the 24 h average relative humidity (%), sea level pressure (mb), temperature (°C), visibility (km), and wind speed (kmph) on the day of atmospheric readings.(6)

9.2.4 POINT SOURCE PB

Under section 313 of the Emergency Planning and Community Right-to-Know Act (EPCRA), firms that release, transfer, or dispose of listed toxins are required to submit annual reports to the EPA detailing quantities of toxins emitted. Data are published under the Toxic Release Inventory (TRI) system.(25) Over the period of observation, 2001 to 2009, a total of 22 Pb-emitting facilities operated in Detroit. In analyses that follow, we estimate whether the presence of a point source polluter of Pb in a child's residential zip code predicts BLL outcomes.

9.2.5 EMPIRICAL STRATEGY

First, we analyze the extent to which daily variation in atmospheric Pb is explained by atmospheric soil.(6) The expectation is that atmospheric Pb and soil are statistically correlated, and that weather-adjusted Pb and soil concentrations in the atmosphere have distinct seasonality, rising and falling simultaneously over the calendar year. Insofar as atmospheric concentrations of soil and Pb move together seasonally, the first link in our soil → air dust → child pathway for Pb exposure is statistically corroborated. We use a least-squares regression procedure to examine the association between atmospheric soil and Pb. Formally, letting y_t denote atmospheric Pb in Detroit on day t our regression estimator is modeled as

$$y_t = \beta_0 + \beta_i S_t + \Gamma_i W_t + \varepsilon_t \tag{2}$$

where β_0 is the model constant; S_t is the atmospheric soil reading on day t; W_t is a vector of local weather conditions on the day atmospheric Pb and soil are measured, and ε_t is the error term, with εt IDD $(0, \sigma_y^2)$. After testing whether atmospheric Pb levels in Detroit are statistically associated with atmospheric soil, we analyze the extent to which variation in child BLLs might be explained by atmospheric Pb concentrations, the second link in our hypothesized pathway.

To determine this relationship, we first analyze child blood Pb as a continuous variable (µg/dL). A census tract fixed effects regression procedure was used to analyze child BLLs. Letting y_{ijt} denote the BLL of child i in census tract j in month t the regression estimator is modeled as

$$y_{ijt} = \alpha_j + \beta + \beta_1 L_i + \beta_2 P_j + \Gamma_1 A_i + \Gamma_2 C_i + \Gamma_3 M_i + \Gamma_4 Z_{it} + \varepsilon_t \tag{3}$$

where α_j is the census tract fixed-effect accounting for unobserved heterogeneity at the neighborhood level; L_t is the average monthly weather-adjusted atmospheric Pb level derived from eq 2 (child BLLs are indexed by month, warranting change in the time-step of Pb aerosol data); P_j is a dummy variable equal to 1 if a Pb emitting facility operates in the zip code of a child's residential location; A_i is the age of the child in years; C_i is a dummy variable equal to 1 if the blood draw was capillary; M_i is a dummy variable equal to 1 if the child is male; Z_{it} corresponds to the year the blood draw occurred, and ε_{ij} is the error term, with $\varepsilon_t \sim$ IDD $(0, \sigma_y^2)$.

Second, using a conditional fixed effect logistic regression procedure, we analyze whether or not a child's BLL is ≥ 5 μg/dL (corresponding to the CDCs new reference value) as a function of atmospheric Pb. Letting Y represent the threshold BLL of a child, where Y equals 1 if a child's BLL is ≥ 5 μg/dL, and Y equals 0 if a child's BLL is <5 μg/dL. We specify the following reduced form logistic equation for the probability of threshold exceedance (Y) for child i in census tract j in month t:

$$Prob(Y_{ijt} = I | L_i, P_j, A_i, C_i, M_i, Z_{it}) = \Lambda[\alpha_j + \beta_1 L_i + \beta_2 P_j + \Gamma_1 A_i + \Gamma_2 C_i + \Gamma_3 M_i + \Gamma_4 Z]_{it} \tag{4}$$

where $\Lambda[\cdot]$ is the CDF of the logistic distribution. The definition of other terms carries from eq 3. The theoretical expectation under eqs 3 and 4 is that child BLL outcomes rise with atmospheric Pb.

Statistical models of child BLLs divide children by reported ages in years. Two reasons motivate this decision. First, low gastric exclusion for Pb in children and high dissolution potential of particulates (due to high surface area to mass ratios) are known to elevate BLLs in children in age-dependent ways.(28) Second, based on strong age-related risk factors observed for children,(29) we logically assume that the effects of airborne exposure are best observed in especially young children (ages 0–2) since they are relatively immobile and more insulated from other known sources of Pb (e.g., paint chips, direct interaction with Pb contaminated soils). Therefore, inasmuch as the proposed soil → air dust → child pathway for Pb exposure is a plausible description of contemporary Pb risk, coefficients on atmospheric Pb in eqs 3 and 4 ought to be noticeably higher in children less than 2 years of age.

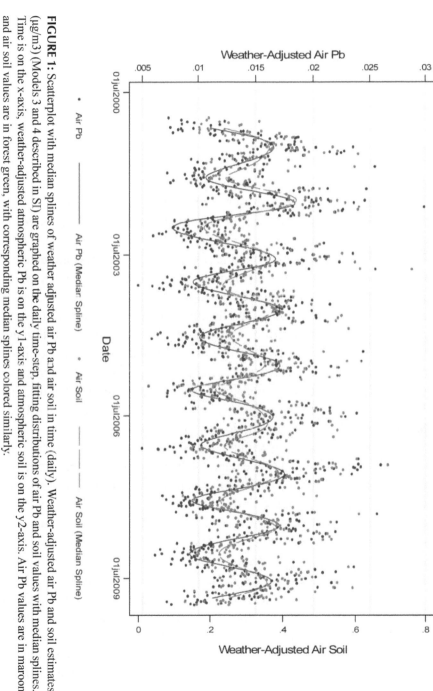

FIGURE 1: Scatterplot with median splines of weather adjusted air Pb and air soil in time (daily). Weather-adjusted air Pb and soil estimates (μg/m3) (Models 3 and 4 described in SI) are graphed on the daily time-step, fitting distributions of air Pb and soil values with median splines. Time is on the x-axis, weather-adjusted atmospheric Pb is on the y1-axis and atmospheric soil is on the y2-axis. Air Pb values are in maroon and air soil values are in forest green, with corresponding median splines colored similarly.

9.3 RESULTS AND DISCUSSION

9.3.1 ATMOSPHERIC SOIL AND PB RELATIONSHIPS

Ordinary least-squares regression models were rendered to predict atmospheric Pb as a function of air soil in Detroit, Michigan from January 2001 to December 2009. Variables are log transformed and elasticities are identified (model details provided in SI). The first model (Model 1) results indicate a 1% increase in atmospheric soil results in a 0.48% (95% CI, 0.38 to 0.58%) increase in atmospheric Pb. The association of air Pb and air soil, adjusting for local weather conditions, is presented as Model 2. Results are similar, with air Pb increasing 0.39% (95% CI, 0.28 to 0.50%) for every 1% increase in atmospheric soil. Standardized betas indicate the atmospheric soil is the strongest predictor of air Pb among variables examined. Note that both atmospheric Pb and soil are similarly sensitive to local weather conditions, rising significantly with average temperature and sea level pressure, and declining significantly with average visibility and wind speed (Models 3 and 4). While traffic-induced resuspension of non-soil particles (derived from wheel weights, brake pads, etc.) is known to be a significant source of atmospheric particulate matter and metals,(30) including a proxy for this variable did not alter the effect observed from atmospheric soil (detailed discussion provided in SI Table SI2).

As shown in Figure 1, weather-adjusted air Pb and soil estimates (derived from Models 3 and 4) have remarkably similar cyclical properties of periodicity, angle function, and amplitude, peaking in the summer/autumn months of June, July, August, and September, and contracting noticeably in the winter months of December and January. The behavior of both Pb and soil aerosol splines parallel the known cyclical behavior of blood Pb outcomes in children observed across many cities and time periods (details of cyclical behavior presented in SI Table SI1 and Figure SI1).(23)

Next, weather-adjusted air Pb and soil estimates are regressed on monthly dummy variables to observe more precisely how much coefficients rise in the months of July, August, and September. Consistent with findings of Laidlaw et al.,(6) compared to the reference of January, atmospheric Pb levels are 35.7% (95% CI, 28.8 to 42.6%) higher in July;

44.8% (95% CI, 38.0 to 51.7%) higher in August; and 40.0% (95% CI, 33.3 to 47.4%) higher in September. Air soil levels also significantly rise in the months of July (57.8%), August (62.2%), and September (54.0%) as compared to the winter month of January (Model 3, SI Table SI4). Overall, atmospheric Pb and soil have remarkably similar seasonal structure, corroborating the claim that a major source of atmospheric Pb is dust that is resuspended from Pb contaminated urban soils.

9.3.2 ATMOSPHERIC PB AND BLOOD PB RELATIONSHIPS

Next, we turn attention to the second link in our hypothesized soil → air dust → child pathway. The natural log of child blood Pb is regressed on air Pb, controlling for child sex, blood draw type, and year of observation (statistical details provided in SI Table SI5). Neighborhood (census tract) fixed effects are incorporated into regression results. Recall that these results are grouped by child age. Log transformation of child blood Pb was necessary given high positive skew (5.87) and kurtosis (164.35), effectively eliminating the skew (0.32) and minimizing kurtosis (2.92). An increase of one standard deviation in weather-adjusted air Pb, 0.0006 $\mu g/m^3$, induces an 8.04% (95% CI, 7.1 to 9.0%) increase in blood Pb for children less than 1 year of age. Air Pb significantly increases blood Pb in all children regardless of age, but the increase in child blood Pb for an equivalent unit of air Pb declines noticeably in age. For instance, while it takes approximately 0.0069 $\mu g/m^3$ of atmospheric Pb to increase child blood Pb by 10% in children 1 year of age, it requires about 3 times the amount (0.023 $\mu g/m^3$) of air Pb to induce a 10% increase in child blood for children 7 years of age.

Interestingly, we also observe an age-specific pattern on differences between males and females in average blood Pb levels (SI Table SI5). No statistically significant difference exists between males and females for children less than 1 year of age. At 1 year of age, a statistically significant difference between boys and girls is apparent and this difference rises incrementally with age. For instance, other things held equal, BLLs in males are only 1.5% higher than females at age 1, but 11.2% higher at age 7. Also worth noting, the effect of a Pb emitting facility in a child's zip code of residence behaves inconsistently by age grouping. For children ages 1–4, we find that residential proximity to a Pb emitting facility increases a child BLL by 2.5–5.5%, depending on the age of the child. For all other ages, proximity to a Pb facility has an effect indistinguishable from chance.

Next, conditional fixed effects odds ratios predicting the expected change in the likelihood of a child's BLL exceeding 5 µg/dL are reported in Table 1. A one standard deviation change in air Pb increases the odds of children less than 1 year of age recording a BLL ≥ 5 µg/dL by a multiplicative factor of 1.32 (95% CI, 1.26 to 1.37). Again, the deleterious effect of air Pb declines with age. By comparison, for 7 year olds, the probability of having a BLL ≥ 5 µg/dL increases 7.8% (95% CI, 2.0 to 13.8%) for an analogous 0.0006 µg/m3 increase in air Pb. As with our linear model, residential proximity to a point source polluter of Pb is a significant predictor of whether a child's BLL is ≥5 µg/dL for children ages 1–3 only.

Figure 2 graphically illustrates the age-dependent association between blood Pb outcomes and air Pb. Overall, results corroborate our intuition that the effects of airborne exposure are most pronounced in younger children (ages 0–2) that are relatively insulated from other sources of Pb (e.g., paint chips, soil Pb).

The age-dependent association between child BLL and air Pb is further visualized by examining the seasonal behavior of age-stratified blood Pb levels and air Pb in time (Figure 3). Three items are notable. First, regardless of age group, average monthly child BLLs in Detroit are definitively seasonal, rising during the summer period and contracting in the winter period. Second, and again regardless of age group, the rise and fall of child BLLs correspond statistically to the behavior of atmospheric Pb. Third, the blood Pb responsiveness to air Pb behaves noticeably differently by age. For children less than 2 years of age, the rise and fall in blood Pb is more congruent with the seasonal behavior of atmospheric Pb. To be more precise, the model fit of average monthly blood Pb as a function of air Pb is substantially higher for children age 0–2 as compared to children 6 years of age and older ($R^2 = 0.706$ vs 0.642). Similarly, a standard deviation rise in atmospheric Pb induces a 0.232 µg/dL (95% CI, 0.203 to 0.260 µg/dL) increase in the monthly average blood Pb of children age 0–2 as compared to a 0.152 µg/dL (95% CI, 0.130 to 0.173 µg/dL) increase in children ≥6 years of age.

TABLE 1: Conditional Fixed Effect Logistic Regression Odds Ratios Predicting Blood Pb (≥ 5 μg/dL) in Detroit Children, 2001–2009[a]

	age 0 odds ratio	age 1 odds ratio	age 2 odds ratio	age 3 odds ratio	age 4 odds ratio	age 5 odds ratio	age 6 odds ratio	age 7 odds ratio	age 8–10 odds ratio
air Pb	1.316***	1.251***	1.177***	1.116***	1.093***	1.128***	1.111***	1.076***	1.089***
	(0.027)	(0.011)	(0.011)	(0.0098)	(0.010)	(0.014)	(0.020)	(0.030)	(0.026)
Pb facility	1.052	1.096**	1.123***	1.141***	1.076	0.893	0.911	0.804	1.057
	(0.091)	(0.043)	(0.050)	(0.048)	(0.046)	(0.054)	(0.077)	(0.100)	(0.109)
capillary draw	1.969***	1.590***	1.678***	1.909***	2.011***	1.823***	1.895***	1.394***	2.415***
	(0.083)	(0.029)	(0.037)	(0.041)	(0.045)	(0.056)	(0.084)	(0.136)	(0.257)
male	1.066	1.052***	1.197***	1.169***	1.225***	1.262***	1.304***	1.427***	1.311***
	(0.043)	(0.017)	(0.022)	(0.021)	(0.022)	(0.033)	(0.047)	(0.082)	(0.064)
year	0.877***	0.860***	0.833***	0.833***	0.827***	0.810***	0.813***	0.800***	0.798***
	(0.007)	(0.003)	(0.003)	(0.003)	(0.003)	(0.004)	(0.006)	(0.010)	(0.008)
N	19,046	75,852	58,322	66,288	66,862	33,878	18,571	8,280	13,122
log likelihood	−7328.2	−42312.1	−33543.5	−37095.8	−35683.7	−17420.0	−8752.6	−3455.4	−5050.8
χ^2	820.8	3684.2	3689.1	4076.6	4132.8	2297.2	1070.8	442.3	658.0
Ncensus tracts	316	370	354	351	350	332	308	287	285

[a]Standard errors in parentheses; ***$p < 0.01$, **$p < 0.05$.

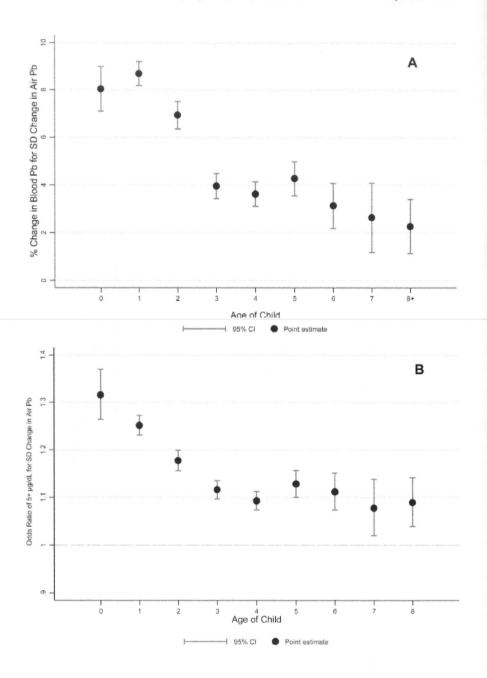

FIGURE 2: Change in blood Pb outcomes (A) and odds ratio of a 5 µg/dL increase in BLL (B) for standard deviation change in air Pb by age (error bars are 95% confidence intervals)

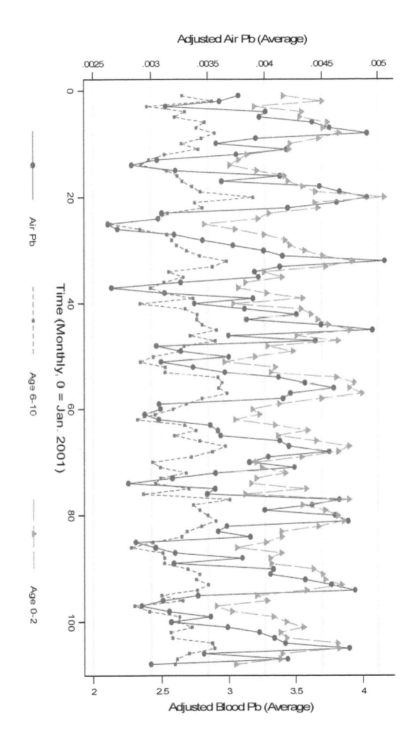

FIGURE 3: Weather-adjusted air Pb (μg/m3) and blood Pb (μg/dL) by age group. Average monthly child blood Pb levels adjusted by local weather conditions, child gender, method of blood draw, and census tract fixed effects; air Pb estimates are from Model 2 (SI Table SI3)

Taken together, regression results on child BLL outcomes as a function of atmospheric Pb described in Table 1 and graphically summarized in Figure 3 corroborate the last link in our hypothesized soil → air dust → child pathway for contemporary Pb risk.

As a logical check on this hypothesized pathway, we end by considering a statistical counterfactual. We use the term counterfactual conventionally, meaning to imagine an outcome with modification of an antecedent. The antecedent that we logically modify is the soil source of air Pb. By regressing child BLL on the average monthly residual of eq 2, we have a test of the effect of air Pb on child BLLs absent soil resuspension. That is, the residual in eq 2 constitutes atmospheric Pb with underlying variation attributable to resuspended soil statistically removed. While the specific content of the residual is unknown, it reflects other unmeasured sources of air Pb; for example, Pb in paint generated by activities such as sanding of home exteriors. Such a test, in effect, interrupts the first link in our causal sequence. Substituting then L_t in eq 4 for ε_t in eq 2, creates then L_t in eq 4 for ε_t in eq 2, creates this counterfactual exercise (results presented in SI Table SI6). With the exception of children age 3, absent soil resuspension, we find that air Pb has an effect on child BLLs indistinguishable from zero. With the soil source of air Pb removed, air Pb has no observable effect on child BLLs—our casual sequence logically disappears.

Overall, empirical analyses corroborate the hypothesized soil → air dust → child pathway for Pb exposure in Detroit children. The data from this study show that daily variation in atmospheric Pb is associated statistically with daily variation in atmospheric soil, with both air Pb and soil showing remarkably similar seasonal properties that match known/observed seasonal variation in child BLLs. In addition, the data demonstrate that air Pb is a significant correlate of child BLLs regardless of age. As expected, and consistent with prior research, the association between child BLLs and air Pb is especially pronounced for children less than 2 years of age.(31) The main exposure mechanism for these young children is likely inadvertent ingestion of fine particulates through hand-to-mouth behavior, exacerbated by poor gastric exclusion for Pb and behavioral patterns that increase surface contact by hand (e.g., crawling). Direct pulmonary uptake remains an untested alternative for some portion of the BLL response. Results from our statistical counterfactual exercise show that absent soil resuspension, the effect of atmospheric Pb on child BLLs is indistinguishable from chance.

The air → dust → child exposure pathway described here may be observed in other urban areas where the legacy Pb deposition in soils remains a critical environmental burden to human health. Our findings suggest that the federal government's continued emphasis on Pb-based paint may be out-of-step (logically) with the evidence presented and an improvement in child health is likely achievable by focusing on the resuspension of soil Pb as a source of exposure. Given that current education has been found to be ineffective in reducing children's exposure to Pb,(15) we recommend that attention be focused on primary prevention of Pb contaminated soils.

REFERENCES

1. Binns, H. J.; Campbell, C.; Brown, M. J.Interpreting and managing blood lead levels of less than 10 μg/dL in children and reducing childhood exposure to lead: Recommendations of the Centers for Disease Control and Prevention advisory committee on childhood lead poisoning prevention Pediatrics 2007, 120 (5) E1285– E1298

2. Canfield, R. L.; Henderson, C. R.; Cory-Slechta, D. A.; Cox, C.; Jusko, T. A.; Lanphear, B. P.Intellectual impairment in children with blood lead concentrations below 10 mu g per deciliter New Eng. J. Med. 2003, 348 (16) 1517– 1526

3. Bellinger, D.; Dietrich, K. N.Low-Level Lead-Exposure and Cognitive Function in Children Pediatr. Ann. 1994, 23 (11) 600– 605

4. Lanphear, B. P.; Hornung, R.; Khoury, J.; Yolton, K.; Baghurstl, P.; Bellinger, D. C.; Canfield, R.; Dietricjh, K.; Bornscheim, R.; Rothenberg, S.; Needleman, H.; Schnaas, L.; Wasserman, G.; Roberst, R.Low-level environmental lead exposure and children's intellectual function: An international pooled analysis Environ. Health Perspect. 2005, 113 (7) 894– 899

5. Jusko, T. A.; Henderson, C. R.; Lanphear, B. P.; Cory-Slechta, D. A.; Parsons, P. J.; Canfield, R. L.Blood lead concentrations < 10 μg/dL and child intelligence at 6 years of age Environ. Health Perspect. 2008, 116 (2) 243– 248

6. Laidlaw, M. A. S.; Zahran, S.; Mielke, H. W.; Taylor, M. P.; Filippelli, G. M.Resuspension of lead contaminated urban soil as a dominant source of atmospheric lead in Birmingham, Chicago, Detroit and Pittsburgh, USA Atmos. Environ. 2012, 49, 302– 310

7. Mielke, H. W.; Gonzales, C. R.; Powell, E.; Jartun, M.; Mielke, P. W.Nonlinear association between soil lead and blood lead of children in metropolitan New Orleans, Louisiana: 2000–2005 Sci. Total Environ. 2007, 388 (1–3) 43– 53

8. Mielke, H. W.; Dugas, D.; Mielke, P. W.; Smith, K. S.; Smith, S. L.; Gonzales, C. R.Associations between soil lead and childhood blood lead in urban New Orleans

and rural Lafourche Parish of Louisiana Environ. Health Perspect. 1997, 105 (9) 950– 954

9. Schroeder, W. H.; Dobson, M.; Kane, D. M.; Johnson, N. D.Toxic Trace Elements Associated with Airborne Particulate Matter: A Review JAPCA 1987, 37 (11) 1267– 1285

10. Zahran, S.; Mielke, H. W.; Gonzales, C. R.; Powell, E. T.; Weiler, S.New Orleans before and after Hurricanes Katrina/Rita: A quasi-experiment of the association between soil lead and children's blood lead Environ. Sci. Technol. 2010, 44 (12) 4433– 40[ACS Full Text ACS Full Text],

11. Hayes, E. B.; McElvaine, M. D.; Orbach, H. G.; Fernandez, A. M.; Lyne, S.; Matte, T. D.Long-term trends in blood lead levels among children in Chicago-Relationship to air lead levels Pediatrics 1994, 93 (2) 195– 200

12. Murray, K. S.; Rogers, D. T.; Kaufman, M. M.Heavy Metals in an Urban Watershed in Southeastern Michigan J. Environ. Qual. 2004, 33 (1) 163– 172

13. Wendland-Bowyer, W.Free Press Soil Study: Samples Offer Cross-Section of Contamination. Detroit Free Press, 2003, January 23, p 10A.

14. Bickel, M. J.Spatial and Temporal Relationships Between Blood Lead and Soil Lead Concentrations in Detroit, Michigan. Masters Thesis, Wayne State University, Detroit, MI, 2010.

15. Yeoh, B.; Woolfenden, S.; Lanphear, B.; Ridley, G. F.; Livingstone, N.Household interventions for preventing domestic Pb exposure in children. Cochrane Database of Systematic Reviews, 2012. 4, Art. CD006047. DOI: 10.1002/14651858.CD006047. pub3.

16. Green, N. A.; Morris, V. R.Assessment of public health risks associated with atmospheric exposure to PM2.5 in Washington, DC, USA Int. J. Environ. Res.Public Health 2006, 3, 86– 97

17. Melaku, S.; Morris, V.; Raghavan, D.; Hosten, C.Seasonal variation of heavy metals in ambient air and precipitation at a single site in Washington, DC Environ. Pollut. 2008, 155 (1) 88– 98

18. USEPA. Seasonal Rhythms of BLL Levels: Boston, 1979–1983: Final Report; EPA 747-R-94-003; U.S. Environmental Protection Agency: Washington, DC, 1995.

19. USEPA. Seasonal Trends in BLL Levels in Milwaukee: Statistical Methodology; EPA 747-R-95-010; U.S. Environmental Protection Agency: Washington, DC, 1996.

20. Billick, I. H.; Curran, A. S.; Shier, D. R.Analysis of pediatric blood lead levels in New York City for 1970–1976 Environ. Health Perspect. 1979, 31, 83– 90

21. New Jersey Department of Health (NJDH) and Senior Services Consumer and Environmental Health Services Environmental Public Health Tracking Project. Analysis of Risk Factors for Elevated Levels of Blood Lead in New Jersey Children, 2000– 2004. Demonstration Project on Geographic Patterns of Childhood Blood Lead and Environmental Factors in New Jersey; Program 03074; Environmental and Health Effects Tracking, National Center for Environmental Health Centers for Disease Control and Prevention (CDC), 2007. http://www.state.nj.us/health/eohs/regional_ state/air/analy_risk_pb_children.pdf.

22. Paode, R. D.; Sofuoglu, S. C.; Sivadechathep, J.; Noll, K. E.; Holsen, T. M.; Keeler, G. J.Dry deposition fluxes and mass size distributions of Pb, Cu, and Zn measured

in southern Lake Michigan during AEOLOS Environ. Sci. Technol. 1998, 32 (11) 1629– 1635

23. Laidlaw, M. A. S.; Mielke, H. W.; Filippelli, G. M.; Johnson, D. L.; Gonzales, C. R.Seasonality and children's blood lead levels: Developing a predictive model using climatic variables and blood lead data from Indianapolis, Indiana, Syracuse, New York, and New Orleans, Louisiana (USA) Environ. Health Perspect. 2005, 113 (6) 793– 800

24. Interagency Monitoring of Protected Visual Environments. http://vista.cira.colostate.edu/improve/ (accessed 9 April 2012) .

25. United States Environmental Protection Agency (USEPA). Toxic Release Inventory (TRI) Program. http://www.epa.gov/tri/ (accessed 9 July 2012) .

26. Zahran, S.; Mielke, H. W.; Weiler, S.; Gonzales, C. R.Nonlinear Associations between Blood Lead in Children, Age of Child, and Quantity of Soil Lead in Metropolitan New Orleans Sci. Total Environ. 2011, 409 (7) 1211– 1218

27. IMPROVE soil estimation calculation. http://vista.cira.colostate.edu/improve/Publications/GrayLit/023_SoilEquation/Soil_Eq_Evaluation.pdf (accessed 9 April 2012)

28. Juhasz, A. L.; Weber, J.; Smith, E.Impact of soil particle size and bioaccessibility on children and adult lead exposure in peri-urban contaminated soils J. Hazard. Mater. 2010, 186 (2–3) 1870– 1879

29. Walter, S. D.; Yankel, A. J.; von Lindern, I. H.Age-specific risk factors for lead absorption in children Arch. Environ. Health 1980, 35 (1) 53– 58

30. Harrison, R. M.; Jones, A. M.; Gietl, J.; Yin, J.; Green, D. C.Estimation of the Contributions of Brake Dust, Tire Wear, and Resuspension to Nonexhaust Traffic Particles Derived from Atmospheric Measurements Environ. Sci. Technol. 2012, 46 (12) 6523– 6529

31. Pino, P.; Walter, T.; Oyarzun, M. J.; Burden, M. J.; Lozoff, B.Rapid drop in infant blood lead levels during the transition to unleaded gasoline use in Santiago, Chile Arch. Environ. Health 2004, 59 (4) 182– 187.

There are several supplemental files that are not available in this version of the article. To view this additional information, please use the citation on the first page of this chapter.

PART V

FOOD AND AGRICULTURE EXPOSURES

CHAPTER 10

NEURODEVELOPMENTAL DISORDERS AND PRENATAL RESIDENTIAL PROXIMITY TO AGRICULTURAL PESTICIDES: THE CHARGE STUDY

JANIE F. SHELTON, ESTELLA M. GERAGHTY, DANIEL J. TANCREDI, LORA D. DELWICHE, REBECCA J. SCHMIDT, BEATE RITZ, ROBIN L. HANSEN, AND IRVA HERTZ-PICCIOTTO

10.1 INTRODUCTION

California is the top agricultural producing state in the nation, grossing 38 billion dollars in revenue from farm crops in 2010 (CDFA 2010). Each year approximately 200 million pounds of active pesticide ingredients are applied throughout the state (CDPR 2011). While pesticides are critical for the modern agricultural industry, certain commonly used pesticides have been associated with abnormal and impaired neurodevelopment in children (Bouchard et al. 2010; Bouchard et al. 2011; Engel et al. 2007; Eskenazi et al. 2006; Grandjean et al. 2006; Guillette et al. 1998; Rauh et al. 2006; Ribas-Fito et al. 2006; Torres-Sanchez et al. 2007; Young et al.

Used with permission from Environmental Health Perspectives. Shelton JF, Geraghty EM, Tancredi DJ, Delwiche LD, Schmidt RJ, Ritz B, Hansen RL, and Hertz-Picciotto I. Neurodevelopmental Disorders and Prenatal Residential Proximity to Agricultural Pesticides: The CHARGE Study. Environmental Health Perspectives, *(2014), http://dx.doi.org/10.1289/ehp.1307044.*

2005). In addition, specific associations have been reported between agricultural pesticides and autism spectrum disorders (ASD) (Roberts et al. 2007) and the broader diagnostic category under which autism falls, the pervasive developmental disorders (Eskenazi et al. 2007).

Developmental delay (DD) refers to young children who experience significant delays reaching milestones in relation to cognitive or adaptive development. Adaptive skills include communication, self-care, social relationships, and/or motor skills. In the U.S., DD affects approximately 3.9% of all children ages 3-10 years, and is approximately 1.7 times more common among boys than girls (Boyle et al. 2011).

Autism is a developmental disorder with symptoms appearing by age three. Specific deficits occur in domains of social interaction and language, and individuals show restricted and repetitive behaviors, activities, or movements (DSM-IV 2000). The autism spectrum disorders (ASDs) represent lower severity, usually with regard to language ability. ASDs affect boys 4-5 times more than girls and the Centers for Disease Control and Prevention recently estimated a prevalence of 1.1% among children 8 years of age, a 78% increase since their 2007 estimate (CDC 2012). Available evidence suggests that causes of both ASD and DD are heterogeneous, and that environmental factors can contribute strongly to risk (Hallmayer et al. 2011; Mendola et al. 2002).

The majority of pesticides sold in the U.S. are neurotoxic and operate through one of three primary mechanisms; 1) inhibition of acetylcholinesterase (AChE), 2) voltage-gated sodium channel disruption, and/or 3) inhibition of gamma-Aminobutyric acid (GABA) (Casida 2009). AChE primarily functions as an inhibitory neurotransmitter, but also has critical roles in the development of learning, cognition, and memory. GABA is also an inhibitory neurotransmitter, and necessary for development and maintenance of neuronal transmission.

Though limited research has assessed in utero exposures to pesticides, animal models (rats) of early exposure to organophosphates showed more severe neurodevelopmental effects for males than females (Levin et al. 2001; Levin et al. 2010). Based on previously published epidemiology or mechanistic considerations, we selected the following pesticide families to investigate for this analysis: organophosphates, carbamates, organochlorines, and pyrethroids. Potential mechanisms linking these select pesticide groups to autism pathophysiology were recently reviewed (Shelton et al. 2012).

The aim of this paper is to explore the relationship between agricultural pesticide applications and neurodevelopmental outcomes by 1) assessing the gestational exposure during pregnancy to CHARGE study mothers, 2) testing the hypothesis that children with ASD or DD had higher risk of exposure in utero than typically developing children, and 3) evaluating specific windows of vulnerability during gestation. Because of the well-defined case and control populations in the CHARGE study, and comprehensive availability of potential confounders, this analysis serves as exploratory research to identify environmental risk factors for ASD and DD, and contributes to a broader understanding of the potential risks to neurodevelopment from agricultural pesticides in a diverse population of California residents.

10.2 METHODS

10.2.1 STUDY DESIGN

The Childhood Autism Risks from Genes and Environment (CHARGE) study is an ongoing California population-based case-control study which aims to uncover a broad array of factors contributing to autism and developmental delay (Hertz-Picciotto et al. 2006). Since 2003, the CHARGE study has enrolled over 1,600 participants whose parents answer extensive questionnaires regarding environmental exposures including their place of residence during pregnancy. Here we report on autism spectrum disorder (ASD) and developmental delay (DD), in relation to gestational residential proximity to agricultural pesticide applications. The group of children with ASD includes approximately two-thirds with a diagnosis of full syndrome autism or autistic disorder (68%) and one third with a diagnosis of an autism spectrum disorder (32%).

Cases are recruited from children diagnosed with full syndrome ASD or DD in one of the regional centers of the California Department of Developmental Services (DDS). Eligibility in the DDS system does not depend on citizenship or financial status, and is widely used across socioeconomic levels and racial/ethnic groups. It is estimated that 75-80% of the total population of children with an autism diagnosis are enrolled in the system (Croen et al. 2002). In addition to recruitment through the regional centers, some CHARGE participants are also recruited through referrals from other clinics, self-referral, or general outreach. The referents are recruited from the general population (GP) identified through California birth records, and are frequency matched to the autism case population on gender, age, and the catchment area for the regional they would have gone to, had they been a case. Children are eligible if they are aged 2-5 years, born in California, live with a biological parent who speaks either English or Spanish, and reside in the study catchment area. Currently, the catchment area for the CHARGE study participants consists of a 2-hr drive from the Sacramento area, but previously included participants from Southern California. Early in the study, recruitment in Southern California was terminated due to logistical difficulties that led to lower enrollment of general population controls.

Parents of children coming into the study with a previous diagnosis of ASD are administered the Autism Diagnostic Interview-Revised (ADI-R), surveyed regarding a wide range of environmental exposures, and asked to report all addresses where they lived from three months before conception to the time of the interview.

Participating children are administered the Autism Diagnostic Observation Schedule (ADOS), and combined with the ADI-R, is used to either confirm their diagnosis or re-classify them for purposes of our study. To rule out ASD, children who enter the study without an ASD diagnosis (from the DD or GP groups) are given the Social Communications Questionnaire (SCQ) (Rutter et al. 2003). Children with a previous diagnosis of DD are evaluated on both the Mullen Scales of Early Learning (MSEL) (Mullen 1995) and Vineland Adaptive Behavioral Scale (VABS) (Sparrow 2005). DD is confirmed if they scored 15 or above on the SCQ and at or below two standard deviations lower than the mean (<70) on the composite scores of MSEL and VABS. Those meeting criteria for one test, scoring <77 on the other, and not qualifying for ASD, are classified as atypical and combined with the DD group (25 out of the 168) for this analysis. For this sample, of those that entered the study as typically developing, 26 were reclassified with DD, and 2 with ASD. Of those who entered as DD, 36 were reclassified with ASD. Only cases with completed diagnostic testing were included in the analysis presented here. Additional details on CHARGE study protocols were published elsewhere (Hertz-Picciotto et al. 2006).

This study was approved by the institutional review boards for the State of California and the University of California. Written informed consent is obtained by the parent or guardian before collection of any data.

10.2.2 ESTIMATION OF PESTICIDE EXPOSURES

Since 1990, California has required commercial application of agricultural pesticides to be reported to the California Department of Pesticide Regulation (CDPR), which makes data publically available in the form of the annual Pesticide Use Report (PUR). As described by CDPR, the pesticide use report data includes "...pesticide applications to parks, golf courses, cemeteries, rangeland, pastures, and along roadside and railroad rights-of-way. In addition, all postharvest pesticide treatments of agricultural commodities must be reported along with all pesticide treatments in poultry and fish production as well as some livestock applications. The primary exceptions to the reporting requirements are home-and-garden use and most industrial and institutional uses."(California Department of Pesticide Regulation 2014).

The PUR database includes all commercial applications at the county level, requiring spatially explicit (latitude and longitude) reporting for commercial agricultural applications. The PUR database then compiles agricultural pesticide applications throughout the state by square-mile areas ($1.0m^2$ or $2.6km^2$) set by the U.S. Geological Survey, referred to as a meridian-township-range-section (MTRS). The amount of chemical applied is assigned to an MTRS by date, in pounds (each pound is 0.45 kg) of active ingredient only, excluding synergists and other compounds in the formulation. Mapping software (ArcGIS v10.0, Esri, Redlands, CA) was used to create a geographic centroid (center-most point in the square mile) for each MTRS for use in this analysis.

From the CHARGE questionnaire administered to the parent, residential addresses were collected and assigned for each day of the pre-conception and pregnancy periods, beginning 3 months prior to conception and ending with delivery, thereby accounting for participants who changed residences during that time. Addresses were manually cleaned for spelling errors and standardized in Zip+4 software (http://www.semaphorecorp. com/). Of the 1,043 diagnostically evaluated participants at the time of this study, 983 had given address data for the time period of interest. Overall, 99% of addresses (970 participants with 1,319 unique addresses) were successfully geocoded to obtain a longitude and latitude with a match rate of at least 80% in ArcMap (ArcGIS v10.0, Esri, Redlands, CA) using the U.S. Rooftop search algorithm. Unmatched addresses (n=5) or ties (n=10) were manually matched to the most likely address.

Next, a spatial model was developed in ArcMap, which created three buffers of varying sizes around each residence with radii of 1.25km, 1.5km, and 1.75km. Where the buffer intersected a centroid (or multiple centroids), the MTRS corresponding with that centroid was assigned to that residence, and subsequently, pesticides applied in that MTRS (or multiple MTRS's) were considered exposures for that mother with the timing based on linking the date of application to the dates of her pregnancy. Each pregnancy therefore was assigned an exposure profile corresponding applications made to the MTRS nearest her home, and days of her pregnancy on which those applications occurred (for a visual representation of the exposure model, see Supplemental Material, Figure S1).

We classified chemicals in the PUR according to chemical structure as members of the organophosphate, carbamate, pyrethroid, or organochlorine classes of pesticides. Sub-classes of pyrethroids were categorized as type 1 and type 2 because they induce distinct behavioral effects in animal studies (ATSDR 2003; Breckenridge et al. 2009). In addition, chlorpyrifos, an organophosphate widely used in agriculture, was explored independently due to previous research that associated higher levels of prenatal exposure with diminished psychomotor and mental development in children at 3 years of age (Rauh et al. 2006).

10.2.3 STATISTICAL ANALYSIS

Most homes (approximately 70%) received estimated agricultural pesticide exposure values of 0 because there was no pesticide applied within the buffer zone. For ease of interpretation, we created, for each time period, binary (1="exposed" vs. 0="not exposed") indicators as independent variables. Multinomial (polytomous) multivariate logistic regression modeling with survey weights was used to estimate the association of prenatal residential proximity to applied pesticides with a binary exposure variable (1=exposed 0=unexposed) and a 3-level case status outcome (ASD / DD / TD), using TD children as the reference group. Because it was the only chemical evaluated independently as opposed to an aggregated class of chemicals with varying toxicity levels, chlorpyrifos (an organophosphate) was evaluated both as a dichotomous (any exposure within the buffer area vs. none) and as a continuous variable (untransformed, per 100lbs). Separate models were run for each time period, for each pesticide class of interest, and for alternative residential buffer radii.

Potential confounders were first identified as variables that 1) may influence ones exposure to pesticides, and 2) variables which are known to influence the risk of ASD or DD, with no requirement for statistical significance of the univariate association with either the exposure or outcome, but rather an initial evaluation of the relationship between those variables. Formal confounder identification and inclusion was assessed using the combined directed acyclic graph (DAG) and change-in-estimate (in this case, a 10% change in the beta of the primary exposure variable in the regression model) criteria (Weng et al. 2009). The DAG was used to establish which variables could potentially confound the associations between ASD or DD and exposure to agricultural pesticides, and the change in estimate criteria was then used to exclude inclusion of those variables that induced minimal (less than 10%) change in the beta estimate. All other variables which were identified as confounders and met the criteria of a 10% or greater change in the beta were included in the final models.

During model selection, the joint versus independent effects of two classes of pesticides was tested (e.g. pyrethroids and organophosphates) in models which contained each independent variable (dichotomous) for the two pesticides and an interaction variable of those two dichotomous variables. We also explored the possibility that another pesticide was responsible for the observed association due to correlation between pesticides (i.e. if one class is applied, another is more likely to be applied in that same buffer zone) by treating other classes of pesticides as potential confounders.

Final models were adjusted for paternal education (categorical), home ownership (binary), maternal place of birth (US, Mexico, or outside of the US and Mexico), child race/ethnicity (white, Hispanic, other), maternal prenatal vitamin intake (binary taken during the three months prior to pregnancy through the first month), and year of birth (continuous). Prenatal vitamin consumption in this time window was found in previous work to have an inverse association with ASD, meaning that early prenatal vitamin intake may confer a lower risk of ASD (Schmidt et al. 2011). Other potential confounders explored but found not to satisfy criteria for confounding based on inclusion in the DAG or the change in estimate criterion were: distance from a major freeway, maternal major metabolic disorders (diabetes, hypertension, and obesity), gestational age (days), latitude of residence, type of insurance used to pay for the delivery (public vs. private), maternal age, paternal age, and season of conception. Maternal age, while a known risk for ASD, does not differ significantly between cases and controls in the CHARGE study because the participating mothers of TD children are older than the general population (see Table 1).

All statistical analyses were conducted using SAS software (version 9.3, Cary, North Carolina). Odds ratios and confidence intervals were estimated using multinomial (polytomous) logistic regression models (PROC SURVEYLOGISTIC) with "survey" weights. Frequency matching factors (regional center, age, and sex of child) were included to adjust for sampling strata using a STRATA statement.

The weights we used in exposure frequency and multinomial models adjusted for differential probabilities of enrollment in the study by case groups (ASD, DD and general population controls) and by social and demographic factors (child race/ethnicity, maternal age, maternal education, insurance payment type at birth, regional center, parity, and maternal birth place) that influence voluntary participation in a case-control study. These weights represent the inverse of the probability of participation, within case and demographically defined groups. Thus, the weighted frequency distributions and regression models more accurately represent findings generalizable to the broader recruitment pools from which participants were drawn.

10.3 RESULTS

During pregnancy, residences of the CHARGE study participants were distributed broadly throughout California, with the greatest concentrations in Sacramento Valley, followed by the San Francisco Bay Area, and Los Angeles. One third lived within 1.5km of an agricultural pesticide application from one of the 4 pesticide classes evaluated. ASD and TD groups had similar socio-demographic profiles, with some variation by regional center, prenatal vitamin intake, and maternal place of birth and more ASD cases were recruited earlier in the study than DD or TD children (Table 1). As described in the methods section, early in the study, challenges were encountered recruiting non-ASD participants in Southern California, resulting in a greater proportion of ASD participants relative to TD participants from that regional center. The DD case group, which was not matched, differed from the reference group on many characteristics, including gender, race/ethnicity, maternal birth place, regional center, maternal education, and paternal education and appears to be of substantially lower socioeconomic status than either the ASD or TD groups (Table1). Age of the child at enrollment was similar between the ASD and DD groups as compared to the TD groups.

In the CHARGE study population, of the pesticides evaluated, organophosphates were the most commonly applied agricultural pesticide near the home during pregnancy. Within the group exposed to organophosphates within 1.5km of the home, twenty-one unique compounds were identified, the most abundant of which was chlorpyrifos (20.7%), followed by acephate (15.4%), and diazinon (14.5%) (Supplemental Material, Table S1). The second most commonly applied class of pesticides was the pyrethroids, one-quarter of which was esfenvalerate (24%), followed by lamda-cyhalothrin (17.3%), permethrin (16.5%), cypermethrin (12.8%), and tau-fluvalinate (10.5%). Of the carbamates, approximately 80% were methomyl or carbaryl, and of the organochlorines, 60% of all applications were dienochlor. Among those exposed, only one-third were exposed to a single compound over the course of the pregnancy.

In the un-weighted study population, little difference in exposure proportion was apparent, yet once the survey weights were applied, both case populations had higher exposure proportions than the typically developing controls, indicating factors associated with exposure were also associated with study participation (Table 2). Because the study weights reflect the distributions of the three recruitment strata (ASD, DD, and population controls) in the pool from which they were drawn, differences between cases and control participation by regional center catchment area likely accounts for this effect (see Methods section for greater detail on survey weights). For example, DD cases proportionally under-enrolled in the CHARGE study from the Valley Mountain regional center as compared to the recruitment pool. Because the Valley region had the highest proportion of exposed participants, weights that accounted for the discrepancy between the proportions of DD cases enrolled from the Valley region would more accurately represent the population distribution of cases and controls.

By pounds applied, the amount of pyrethroids and organophosphates (continuous, un-weighted) within 1.5km of the home were strongly correlated with each other ($\rho=0.74$, $p<0.0001$) and to a lesser extent organophosphates with carbamates ($\rho=0.45$, $p=0.01$) and carbamates with pyrethroids ($\rho=0.44$, $p<0.0001$). Due to the low prevalence of organochlorines and type 1 pyrethroids, they were excluded from the analyses, and carbamate exposure, while evaluated for pregnancy (any vs. none), was not evaluated by trimester due to small cell sizes of exposed participants. Overall, exposure to pesticides during gestation was slightly more common for male children than female children (31% vs. 26%, $p=0.004$).

For exposure (any vs. none) during pregnancy, children with ASD were 60 per cent more likely to have organophosphates applied nearby the home (1.25km distance, aOR=1.60, 95% CI=1.02-2.51) than mothers of TD children. Children with DD were nearly 150 per cent more likely to have carbamate pesticides applied near the home during pregnancy (1.25km distance, aOR=2.48, 95% CI=1.04-5.91). Both of these associations lessened as the buffer size grew larger (Tables 3 and 4), lending support to an exposure-response gradient Examining specific gestational time windows, associations with pesticide applications of organophosphates and pyrethroids suggested an association between 2nd and 3rd trimester exposure to organophosphates and ASD, and pre-conception and 3rd trimester pyrethroid exposure (Table 3). While those time periods describe the statistically significant associations, many of the effect estimates tended away from the null, which indicates a lack of precision in the specificity of any one time period and compound presented here.

For DD, the sample size only permitted temporal associations to be evaluated for organophosphates and pyrethroids, which were mostly higher than 1 (the null value), but only one statistically significant association was detected for 3rd trimester pyrethroid applications. In general, likely due to a smaller sample of DD cases exposed to agricultural pesticides, the estimates had a lower level of precision than the ASD case group. In addition, although carbamates were associated with DD for applications during pregnancy, the sample of exposed cases was too small to evaluate by trimester (Table 4).

For models evaluating the exposure to chlorpyrifos as a continuous variable with all other covariates remaining the same as above models, each 100 pound (45.4kg) increase in the amount applied over the course of pregnancy (within 1.5km of the home) was associated with a 14% higher prevalence of ASD (aOR=1.14, 95% CI= 1.0, 1.32), but no association was detected with DD. Because aggregate classes of chemical do not have a uniform toxicity, we did not examine the pounds of classes (e.g. organophosphates) of chemicals as a continuous variable because compounds with a higher toxicity compounds may be applied in lower volumes.

The role of simultaneous exposure to multiple classes of pesticides was evaluated in post-hoc analyses. First, we evaluated combined categories of organophosphates and pyrethroids, organophosphates and carbamates, and pyrethroids or carbamates as a 3-level variable (0=unexposed, 1=exposed to one or the other, and 2=exposed to both). However, effects from multiple exposures were not found to be higher than the observations of the individual classes of pesticides. Second, we adjusted models of one pesticide for the other. In models for organophosphates, adjusting for pyrethroids attenuated the 3rd trimester association with ASD slightly, but not substantially (less than 10% change in β estimate) (data not shown). In additional analyses, we evaluated the sensitivity of the estimates to the choice of buffer size, using 4 additional sizes between 1 and 2km: results and interpretation remained stable (data not shown).

10.4 DISCUSSION

Applications of two of the most common agricultural pesticides (organophosphates and pyrethroids) nearby the home may increase the prevalence of ASD. Specifically, we observed positive associations between ASD and prenatal residential proximity to organophosphate pesticides in the 2nd (for chlorpyrifos) and 3rd trimesters (organophosphates overall), and pyrethroids in the three months prior to conception and in the 3rd trimester. Our findings relating agricultural pesticides to DD were less robust, but were suggestive of an associated with applications of carbamates during pregnancy nearby the home. Because pesticide exposure is correlated in space and time, differences in time-windows of vulnerability, if they exist, may be difficult to detect, and variation in associations according to time window of exposure may not represent causal variation.

These findings support the results of two previous studies linking ASD to gestational agricultural pesticide exposure. Using data from the California Department of Developmental Services and California Birth Records, Roberts et al. (2007) conducted a case-control study of 465 cases of autism and 6,975 controls. Although their main finding was an association between ASD and residential proximity to organochlorine compound applications (which we could not evaluate due to low exposure prevalence of this chemical class), they also reported associations with gestational exposures to organophosphates ($\beta=0.462$, p-value 0.042 (confidence interval not reported) and bifenthrin ($\beta=1.57$, p-value=0.049 (confidence interval not reported)), a pyrethroid pesticide (Roberts et al. 2007). Eskenazi et al. (2007) found a relationship between symptoms of pervasive developmental disorder (PDD) and prenatal urinary metabolites of organophosphates in a cohort study (named CHAMACOS) of mothers living in the Salinas valley. Each ten-fold increase in these metabolites doubled the odds (OR=2.3, p=0.05) of PDD at two years of age; postnatal concentrations showed some association as well (OR=1.7, p=0.04) (Eskenazi et al. 2007). Several studies have also reported evidence of an interaction between organophosphate exposure and polymorphisms for the PON1 gene, which codes for the enzyme paroxonase 1, in relation to neurodevelopment (Costa et al. 2005; D'Amelio et al. 2005; Furlong et al. 2005; Lee et al. 2013).

With regard to DD, several studies have reported associations of pesticide exposures with continuous scores on specific cognitive tests. For example, in a cross-sectional study of 72 children under 9 years of age in Ecuador, those prenatally exposed to pesticides as assessed by maternal occupation in the floriculture industry during pregnancy, performed worse on the Stanford-Binet copying test than did children whose mothers did not work in floriculture during pregnancy (Grandjean et al. 2006). In another study of maternal occupation in the flower industry, exposed children performed worse on tests of communication, visual acuity, and fine motor skills, with delays of 1.5 to 2 years in reaching normal developmental milestones (Handal et al. 2008). In the CHAMACOS cohort, organophosphate urinary metabolites from the 1st and 2nd halves of pregnancy were associated with an average deficit of 7.0 IQ points, comparing the highest quintile to the lowest (Bouchard et al. 2011). A study of inner-city at 3 years of age found that those with the highest (versus lowest) umbilical cord concentrations of chlorpyrifos were five times more likely to have delayed psychomotor development and 2.4 times more likely to have delayed mental development as assessed by cut-off values of continuous scores on the Bayley Scales of Infant Development-II (Rauh et al. 2006).

Strengths of this study include well-defined case and control populations confirmed by standardized diagnostic instruments, extensive information on covariates, and a thorough confounder identification and control strategy. Because children can overcome developmental delay, or may move in or out of the ASD case definition over time, diagnostic confirmation at enrollment minimized outcome misclassification. Further, collection of information on all addresses during pregnancy likely reduced exposure misclassification, as 20% of the population had moved at least once during pregnancy.

Several limitations to this study were unavoidable in the exposure assessment, potentially producing misclassification. Primarily, our exposure estimation approach does not encompass all potential sources of exposure to each of these compounds: among them external non-agricultural sources (e.g. institutional use, such as around schools); residential indoor use; professional pesticide application in or around the home for gardening, landscaping or other pest control; as well as dietary sources (Morgan 2012). Other sources of potential error include errors in reporting to the

Pesticide Use Report data base, the assumption of homogeneity of exposure within each buffer, and potential geo-coding errors. Seasonal variation and address changes mid-pregnancy were accounted for by assigning
an address to each day instead of one address for the individual, but information on hours spent in the home or elsewhere was not available.

Utilization of the PUR data has been refined by some researchers who
have enhanced the 1 square mile resolution of the PUR data by incorporating land use data (Nuckols et al. 2007; Rull and Ritz 2003). This approach demonstrates higher correlation of PUR-based exposure estimates
with in-home carpet dust pesticide concentrations than the PUR data alone
(Gunier et al. 2011). In our case, land use reports were not available for
about half the CHARGE study counties; given an already low prevalence
of exposure, the loss of power by excluding those counties would have
outweighed any benefit of increased specificity in exposure estimates from
land-use data.

Although organophosphate use drastically increased between the
1960's through the late 1990's (USDA 2006), over the past-decade, use
has been declining (US EPA 2011). For indoor use, chlorpyrifos has largely been replaced with pyrethroids (Williams et al. 2008), but research indicates pyrethroids may not necessarily be safer. In an in vitro study comparing the toxicity of a common pyrethroid, cyfluthrin, to chlorpyrifos,
at the same doses cyfluthrin induced either an equivalent or higher toxic
effect on the growth, survival and function of primary fetal human astrocytes, and induced inflammatory action of astrocytes which can mediate
neurotoxicity (Mense et al. 2006). In another in vitro study comparing
the neurotoxicity of fipronil to chlorpyrifos, fipronil induced more oxidative stress and resulted in lower cell counts for non-differentiated PC12
cells than chlorpyrifos, and disrupted cell development at lower thresholds, leading authors to conclude fipronil was in fact more detrimental to
neuronal cell development than chlorpyrifos (Lassiter et al. 2009). While
further studies are underway, because of the observed associations in humans and direct effects on neurodevelopmental toxicity in animal studies,
caution is warranted for women to avoid direct contact with pesticides
during pregnancy.

10.5 CONCLUSIONS

Children of mothers who live near agricultural areas, or who are otherwise exposed to organophosphate, pyrethroid, or carbamate pesticides during gestation may be at increased risk for neurodevelopmental disorders. Further research on gene-by-environment interactions may reveal vulnerable sub-populations.

REFERENCES

1. ATSDR. 2003. Toxicological Profile for Pyrethrins and Pyrethroids. Available: http://www.atsdr.cdc.gov/toxprofiles/tp155.pdf
2. Bouchard MF, Bellinger DC, Wright RO, Weisskopf MG. 2010. Attention-deficit/hyperactivity disorder and urinary metabolites of organophosphate pesticides. Pediatrics 125(6):e1270-1277.
3. Bouchard MF, Chevrier J, Harley KG, Kogut K, Vedar M, Calderon N, et al. 2011. Prenatal exposure to organophosphate pesticides and IQ in 7-year-old children. Environ Health Perspect 119(8):1189-1195.
4. Boyle CA, Boulet S, Schieve LA, Cohen RA, Blumberg SJ, Yeargin-Allsopp M, et al. 2011. Trends in the prevalence of developmental disabilities in US children, 1997-2008. Pediatrics 127(6):1034-1042.
5. Breckenridge CB, Holden L, Sturgess N, Weiner M, Sheets L, Sargent D, et al. 2009. Evidence for a separate mechanism of toxicity for the Type I and the Type II pyrethroid insecticides. Neurotoxicology 30 Suppl 1:S17-31.
6. CDPR (California Department of Pesticide Regulation). 2014. Pesticide Use Reporting (PUR). Available:http://www.cdpr.ca.gov/docs/pur/purmain.htm [accessed 22 May 2014].
7. Casida JE. 2009. Pest toxicology: the primary mechanisms of pesticide action. Chem Res Toxicol 22(4):609-619.
8. CDC (Centers for Disease Control and Prevention). 2012. Prevalence of autism spectrum disorders - autism and developmental disabilities monitoring network, 14 sites, United States, 2008. (MMWR Surveill Summ). Available: http://www.cdc.gov/mmwr/preview/mmwrhtml/ss6103a1.htm [accessed 27 May 2014]
9. CDFA (California Department of Food and Agriculture). 2010. California Agricultural Production Statistics. Available: http://www.cdfa.ca.gov/statistics/ [accessed February 12, 2012].
10. CDPR (California Department of Pesticide Regulation). 2011. Pesticide Use Report. Available:http://www.cdpr.ca.gov/docs/pur/purmain.htm [accessed January 5, 2011].

11. Costa LG, Cole TB, Vitalone A, Furlong CE. 2005. Measurement of paraoxonase (PON1) status as a potential biomarker of susceptibility to organophosphate toxicity. Clin Chim Acta 352(1-2):37-47.

12. Croen LA, Grether JK, Hoogstrate J, Selvin S. 2002. The changing prevalence of autism in California. J Autism Dev Disord 32(3):207-215.

13. D'Amelio M, Ricci I, Sacco R, Liu X, D'Agruma L, Muscarella LA, et al. 2005. Paraoxonase gene variants are associated with autism in North America, but not in Italy: possible regional specificity in gene-environment interactions. Mol Psychiatry 10(11):1006-1016.

14. DSM-IV. 2000. Diagnostic and Statistical Manual of Mental Disorders-IV-TR. Washington, DC: American Psychiatric Association.

15. Engel SM, Berkowitz GS, Barr DB, Teitelbaum SL, Siskind J, Meisel SJ, et al. 2007. Prenatal organophosphate metabolite and organochlorine levels and performance on the Brazelton Neonatal Behavioral Assessment Scale in a multiethnic pregnancy cohort. Am J Epidemiol 165(12):1397-1404.

16. Eskenazi B, Marks AR, Bradman A, Fenster L, Johnson C, Barr DB, et al. 2006. In utero exposure to dichlorodiphenyltrichloroethane (DDT) and dichlorodiphenyl-dichloroethylene (DDE) and neurodevelopment among young Mexican American children. Pediatrics 118(1):233-241.

17. Eskenazi B, Marks AR, Bradman A, Harley K, Barr DB, Johnson C, et al. 2007. Organophosphate pesticide exposure and neurodevelopment in young Mexican-American children. Environ Health Perspect 115(5):792-798.

18. Furlong CE, Cole TB, Jarvik GP, Pettan-Brewer C, Geiss GK, Richter RJ, et al. 2005. Role of paraoxonase (PON1) status in pesticide sensitivity: genetic and temporal determinants. Neurotoxicology 26(4):651-659.

19. Grandjean P, Harari R, Barr DB, Debes F. 2006. Pesticide exposure and stunting as independent predictors of neurobehavioral deficits in Ecuadorian school children. Pediatrics 117(3):e546-556.

20. Guillette EA, Meza MM, Aquilar MG, Soto AD, Garcia IE. 1998. An anthropological approach to the evaluation of preschool children exposed to pesticides in Mexico. Environ Health Perspect 106(6):347-353.

21. Gunier RB, Ward MH, Airola M, Bell EM, Colt J, Nishioka M, et al. 2011. Determinants of agricultural pesticide concentrations in carpet dust. Environ Health Perspect 119(7):970-976.

22. Hallmayer J, Cleveland S, Torres A, Phillips J, Cohen B, Torigoe T, et al. 2011. Genetic heritability and shared environmental factors among twin pairs with autism. Arch Gen Psychiatry 68(11):1095-1102.

23. Handal AJ, Harlow SD, Breilh J, Lozoff B. 2008. Occupational exposure to pesticides during pregnancy and neurobehavioral development of infants and toddlers. Epidemiology 19(6):851-859.

24. Hertz-Picciotto I, Croen LA, Hansen R, Jones CR, van de Water J, Pessah IN. 2006. The CHARGE study: an epidemiologic investigation of genetic and environmental factors contributing to autism. Environ Health Perspect 114(7):1119-1125.

25. Lassiter TL, MacKillop EA, Ryde IT, Seidler FJ, Slotkin TA. 2009. Is fipronil safer than chlorpyrifos? Comparative developmental neurotoxicity modeled in PC12 cells. Brain Res Bull 78(6):313-322.

26. Lee PC, Rhodes SL, Sinsheimer JS, Bronstein J, Ritz B. 2013. Functional para-oxonase 1 variants modify the risk of Parkinson's disease due to organophosphate exposure. Environment international 56C:42-47.

27. Levin ED, Addy N, Nakajima A, Christopher NC, Seidler FJ, Slotkin TA. 2001. Persistent behavioral consequences of neonatal chlorpyrifos exposure in rats. Brain research Developmental brain research 130(1):83-89.

28. Levin ED, Timofeeva OA, Yang L, Petro A, Ryde IT, Wrench N, et al. 2010. Early postnatal parathion exposure in rats causes sex-selective cognitive impairment and neurotransmitter defects which emerge in aging. Behav Brain Res 208(2):319-327.

29. Mendola P, Selevan SG, Gutter S, Rice D. 2002. Environmental factors associated with a spectrum of neurodevelopmental deficits. Mental retardation and developmental disabilities research reviews 8(3):188-197.

30. Mense SM, Sengupta A, Lan C, Zhou M, Bentsman G, Volsky DJ, et al. 2006. The common insecticides cyfluthrin and chlorpyrifos alter the expression of a subset of genes with diverse functions in primary human astrocytes. Toxicol Sci 93(1):125-135.

31. Morgan MK. 2012. Children's Exposures to Pyrethroid Insecticides at Home: A Review of Data Collected in Published Exposure Measurement Studies Conducted in the United States. International journal of environmental research and public health 9(8):2964-2985.

32. Hallmayer J, Cleveland S, Torres A, Phillips J, Cohen B, Torigoe T, et al. 2011. Genetic heritability and shared environmental factors among twin pairs with autism. Arch Gen Psychiatry 68(11):1095-1102.

33. Handal AJ, Harlow SD, Breilh J, Lozoff B. 2008. Occupational exposure to pesticides during pregnancy and neurobehavioral development of infants and toddlers. Epidemiology 19(6):851-859.

34. Hertz-Picciotto I, Croen LA, Hansen R, Jones CR, van de Water J, Pessah IN. 2006. The CHARGE study: an epidemiologic investigation of genetic and environmental factors contributing to autism. Environ Health Perspect 114(7):1119-1125.

35. Lassiter TL, MacKillop EA, Ryde IT, Seidler FJ, Slotkin TA. 2009. Is fipronil safer than chlorpyrifos? Comparative developmental neurotoxicity modeled in PC12 cells. Brain Res Bull 78(6):313-322.

36. Lee PC, Rhodes SL, Sinsheimer JS, Bronstein J, Ritz B. 2013. Functional para-oxonase 1 variants modify the risk of Parkinson's disease due to organophosphate exposure. Environment international 56C:42-47.

37. Levin ED, Addy N, Nakajima A, Christopher NC, Seidler FJ, Slotkin TA. 2001. Persistent behavioral consequences of neonatal chlorpyrifos exposure in rats. Brain research Developmental brain research 130(1):83-89.

38. Levin ED, Timofeeva OA, Yang L, Petro A, Ryde IT, Wrench N, et al. 2010. Early postnatal parathion exposure in rats causes sex-selective cognitive impairment and neurotransmitter defects which emerge in aging. Behav Brain Res 208(2):319-327.

39. Mendola P, Selevan SG, Gutter S, Rice D. 2002. Environmental factors associated with a spectrum of neurodevelopmental deficits. Mental retardation and developmental disabilities research reviews 8(3):188-197.

40. Mense SM, Sengupta A, Lan C, Zhou M, Bentsman G, Volsky DJ, et al. 2006. The common insecticides cyfluthrin and chlorpyrifos alter the expression of a subset of

genes with diverse functions in primary human astrocytes. Toxicol Sci 93(1):125-135.

41. Morgan MK. 2012. Children's Exposures to Pyrethroid Insecticides at Home: A Review of Data Collected in Published Exposure Measurement Studies Conducted in the United States. International journal of environmental research and public health 9(8):2964-2985.

42. US EPA (United States Environmental Protection Agency. 2011. Pesticide Industry Sales and Usage. Available: http://www.epa.gov/opp00001/pestsales/07pestsales/market_estimates2007.pdf [accessed 27 May 2014]

43. USDA (United States Department of Agriculture). 2006. Pest Management Practices. Available: http://www.ers.usda.gov/ersDownloadHandler.ashx?file=/media/873656/pestmgt.pdf [accessed June 2, 2014].

44. Weng HY, Hsueh YH, Messam LL, Hertz-Picciotto I. 2009. Methods of covariate selection: directed acyclic graphs and the change-in-estimate procedure. Am J Epidemiol 169(10):1182-1190.

45. Williams MK, Rundle A, Holmes D, Reyes M, Hoepner LA, Barr DB, et al. 2008. Changes in pest infestation levels, self-reported pesticide use, and permethrin exposure during pregnancy after the 2000-2001 U.S. Environmental Protection Agency restriction of organophosphates. Environ Health Perspect 116(12):1681-1688.

46. Young JG, Eskenazi B, Gladstone EA, Bradman A, Pedersen L, Johnson C, et al. 2005. Association between in utero organophosphate pesticide exposure and abnormal reflexes in neonates. Neurotoxicology 26(2):199-209.

There are several supplemental tables that are not available in this version of the article. To view this additional information, please use the citation on the first page of this chapter.

CHAPTER 11

SEVEN-YEAR NEURODEVELOPMENTAL SCORES AND PRENATAL EXPOSURE TO CHLORPYRIFOS, A COMMON AGRICULTURAL PESTICIDE

VIRGINIA RAUH, SRIKESH ARUNAJADAI, MEGAN HORTON, FREDERICA PERERA, LORI HOEPNER, DANA B. BARR, AND ROBIN WHYATT

Each year, thousands of new chemicals are released in the United States, with very little documentation about potential long-term human health risks (Landrigan et al. 2002). First registered in 1965 for agricultural and pest control purposes, chlorpyrifos (CPF; 0,0-diethyl-0-3,5,6-trichloro-2-pyridyl phosphorothioate) is a broad-spectrum, chlorinated organophosphate (OP) insecticide. Before regulatory action by the U.S. Environmental Protection Agency (EPA) to phase out residential use beginning in 2000, CPF applications were particularly heavy in urban areas, where the exposed populations included pregnant women (Berkowitz et al. 2003; Whyatt et al. 2002, 2003). In a sample of pregnant women in New York City (Perera et al. 2002) detectable levels of CPF were found in 99.7% of personal air samples, 100% of

Used with permission from *Environmental Health Perspectives. Rauh V, Arunajadai S, Horton M, Perera F, Hoepner L, Barr DB, Whyatt R. Seven-Year Neurodevelopmental Scores and Prenatal Exposure to Chlorpyrifos, a Common Agricultural Pesticide.* Environmental Health Perspectives, *119 (2011),* http://dx.doi.org/10.1289/ehp.1003160.

indoor air samples, and 64–70% of blood samples collected from umbilical cord plasma at delivery (Whyatt et al. 2002).

Early concerns about the possible neurotoxicity of OP insecticides for humans derived from rodent studies showing that prenatal and early post-natal exposures to CPF were associated with neurodevelopmental deficits, and these effects have been seen at exposure levels well below the threshold for systemic toxicity caused by cholinesterase inhibition in the brain (e.g., Slotkin and Seidler 2005). Evidence has accumulated over the past decade showing that noncholinergic mechanisms may play a role in the neurotoxic effects of CPF exposure in rodents, involving disruption of neural cell development, neurotransmitter systems (Aldridge et al. 2005; Slotkin 2004), and synaptic formation in different brain regions (Qiao et al. 2003). Such developmental disruptions have been associated with later functional impairments in learning, short-term working memory, and long-term reference memory (Levin et al. 2002).

In humans, OPs have been detected in amnionic fluid (Bradman et al. 2003) and are known to cross the placenta (Richardson 1995; Whyatt et al. 2005), posing a threat to the unborn child during a period of rapid brain development. Using urinary metabolites as the biomarker of exposure, several different birth cohort studies have reported that prenatal maternal nonspecific OP exposure was associated with abnormal neonatal reflexes (Engel et al. 2007; Young et al. 2005), mental deficits and pervasive development disorder at 2 years (Eskenazi et al. 2007), and attention problem behaviors and a composite attention-deficit/hyperactivity disorder indicator at 5 years of age (Marks et al. 2010).

Using a different biomarker of exposure (the parent compound of CPF in umbilical cord plasma), we have previously reported (in the same cohort as the present study) significant associations between prenatal exposure to CPF (> 6.17 pg/g) and reduced birth weight and birth length (Whyatt et al. 2004), increased risk of small size for gestational age (Rauh V, Whyatt R, Perera F, unpublished data), increased risk of mental and motor delay (< 80 points) and 3.5- to 6-point adjusted mean decrements on the 3-year Bayley Scales of Infant Development (Rauh et al. 2006), and evidence of increased problems related to attention, attention deficit hyperactivity disorder, and pervasive developmental disorder as measured by the Child Behavior Checklist at 2–3 years (Rauh et al. 2006). Taken together, these prospective cohort studies show a consistent pattern of early cognitive and behavioral deficits related to prenatal OP exposure, across both agricultural and urban populations, using different biomarkers of prenatal exposure.

We undertook the present study to identify the developmental consequences of prenatal exposure to CPF in a sample of New York City children at 7 years of age. Given the mechanisms proposed in the rodent literature, and early findings from prospective human studies involving nonspecific OP exposures, we hypothesized that prenatal exposure to CPF would be associated with neurodevelopmental deficits persisting into the early school years, when more refined neuropsychological tests are available to identify particular functional impairments.

11.1 MATERIALS AND METHODS

Participants and recruitment. The subjects for this report are participants in an ongoing prospective cohort study (Columbia Center for Children's Environmental Health) of inner-city mothers and their newborn infants (Perera et al. 2002). The cohort study was initiated in 1997 to evaluate the effects of prenatal exposures to ambient pollutants on birth outcomes and neurocognitive development in a cohort of mothers and newborns from low-income communities in New York City. Nonsmoking women (classified by self-report and validated by blood cotinine levels < 15 ng/mL), 18–35 years of age, who self-identified as African American or Dominican and who registered at New York Presbyterian Medical Center or Harlem Hospital prenatal clinics by the 20th week of pregnancy, were approached for consent. Eligible women were free of diabetes, hypertension, known HIV, and documented drug abuse and had resided in the area for at least 1 year. The study was approved by the Institutional Review Board of Columbia University. Informed consent was obtained from all participating mothers, and informed assent was obtained from all children as well, starting at 7 years of age.

Of 725 consenting women, 535 were active participants in the ongoing cohort study at the time of this report, and 265 of their children had reached the age of 7 years with complete data on the following: a) prenatal maternal interview data, b) biomarkers of prenatal CPF exposure level from maternal and/or cord blood samples at delivery, c) postnatal covariates, and d) neurodevelopmental outcomes.

Maternal interview and assessment. A 45-min questionnaire was administered to each woman in her home by a trained bilingual interviewer during the third trimester of pregnancy and annually thereafter. From the interviews and medical records, the following sociodemographic and biomedical variables, among others, were available: race/ethnicity, infant sex, household income, maternal age, maternal completed years of education at child's age 7 years, birth weight, gestational age, and self-reported maternal exposure to environmental tobacco smoke (ETS) during pregnancy.

We measured maternal nonverbal intelligence by the Test of Nonverbal Intelligence, 3rd edition (TONI-3) (Brown et al. 1997), a 15-min language-free measure of general intelligence, administered when the child was 3 years of age. The quality of the care-taking environment was measured by the Home Observation for Measurement of the Environment (HOME) inventory when the child was 3 years of age (Caldwell and Bradley 1979) to assess physical and interactive home characteristics. The mother report version of the Child Behavior Checklist for ages 6–18 years, a well-validated measure of child behavior problems occurring in the preceding 2 months (Achenbach and Rescorla 2001), was administered at 7 years as part of the larger cohort study.

Biological samples and pesticide exposure. A sample of umbilical cord blood (30–60 mL) was collected at delivery, and a sample of maternal blood (30–35 mL) was collected within 2 days postpartum by hospital staff. Portions were sent to the Centers for Disease Control and Prevention (Atlanta, GA) for analysis of CPF in plasma, as well as lead and cotinine, described in detail elsewhere (Perera et al. 2002; Whyatt et al. 2003). Methods for the laboratory assay for CPF, including quality control, reproducibility, and limits of detection (LODs), have also been previously published (Barr et al. 2002). In cases where the umbilical cord blood sample was not collected (12% of subjects), mothers' values were substituted, using a formula previously derived from regression analyses (Whyatt et al. 2005). As previously reported, maternal and umbilical cord blood CPF concentrations were similar (arithmetic means ± SDs of 3.9 ± 4.8 pg/g for maternal blood and 3.7 ± 5.7 pg/g for cord blood) (Whyatt et al. 2005), and CPF levels in paired maternal and umbilical cord plasma samples were highly correlated ($r = 0.76$; $p < 0.001$, Spearman's rank), indicating that CPF was readily transferred from mother to fetus during pregnancy. Prenatal blood lead levels were available for a subset of children ($n = 89$). ETS exposure, measured by maternal self-report, was validated by cotinine levels in umbilical cord blood, as described in detail elsewhere (Rauh et al. 2004). We measured polycyclic aromatic hydrocarbon (PAH) exposure by personal air monitoring during the third trimester, using a previously described method, and excluding poor-quality samples (Perera et al. 2003). As previously described (Perera et al. 2003), we computed a composite log-transformed PAH variable from the eight correlated PAH air concentration measures (r-values ranging from 0.34 to 0.94; all p-values < 0.001 by Spearman's rank).

In the larger cohort study, $> 40\%$ of CPF exposure values for combined maternal and umbilical cord blood samples were below the LOD. Using a method suggested by Richardson and Ciampi (2003), we made a distributional assumption for the exposure variable (log-normal CPF), computed the expected value of the exposure (E) for all nondetects [$E(X/X < LOD)$], and assigned this value to all nondetects.

Measures of neurodevelopment. For the 7-year assessment, we selected the Wechsler Intelligence Scale for Children, 4th edition (WISC-IV), because of its revised structure based on the latest research in neurocognitive models of information processing (Wechsler 2003). The WISC-IV is sensitive to low-dose neurotoxic exposures, as demonstrated by studies of lead toxicity in 6- to 7.5-year-old children (Chiodo et al. 2004; Jusko et al. 2008; Rothenberg and Rothenberg 2005). The instrument measures four areas of mental functioning that are associated with, but distinct from, overall intelligence quotient (IQ) and is sensitive to cognitive deficits related to learning and working memory, which have been linked to CPF exposure in rodent studies (e.g., Levin et al. 2002). Each standardized scale has a mean of 100 and SD of 15. The Verbal Comprehension Index is a measure of verbal concept formation, a good predictor of school readiness (Hecht et al. 2000; Wechsler 2003); the Perceptual Reasoning Index measures nonverbal and fluid reasoning; the Working Memory Index assesses children's ability to memorize new information, hold it in short-term memory, concentrate, and manipulate information; the Processing Speed Index assesses ability to focus attention and quickly scan, discriminate, and sequentially order visual information; and the Full-Scale IQ score combines the four composite indices. The General Ability Index score is a summary score of general intelligence, similar to Full-Scale IQ, but excludes contributions from both Working Memory Index and Processing Speed Index (Wechsler 2003). WISC-IV scores may be influenced by socioeconomic background and/or child behavior problems particularly those related to anxiety (Wechsler 2003).

Data analysis. We conducted all analyses using the statistical program R (R Development Core Team 2010). We treated CPF exposure level (picograms per gram) as a continuous variable. We natural log (ln) transformed the WISC-IV Composite Index scores to stabilize the variance and to improve the linear model fit, based on regression diagnostics. Unadjusted correlation analyses were used to explore associations between CPF exposure and WISC-IV scores. We constructed smoothed cubic splines to explore the shape of the functional relationships between CPF exposure and each of the log-transformed WISC-IV indices. We compared the models in which CPF is entered as a single continuous outcome with those in which CPF is modeled using B-splines, using the Davidson–MacKinnon J-test for comparing nonnested models (Davidson and MacKinnon 1981).

Demographic, biomedical, and chemical exposure variables collected for the larger cohort study were available for possible inclusion in the present analysis. We used two different approaches for covariate selection and model fitting, for the purpose of determining the robustness of our results with respect to alternate methods. Covariates were initially selected based on prior literature and retained in the models if associated with either CPF exposure or the WISC-IV scales ($p < 0.10$ in univariate analyses). Multiple linear regression was used to test the effects of prenatal CPF exposure on each 7-year WISC-IV Index. We examined residuals for normality and homoscedasticity and detected no problems. In addition, we employed the least absolute shrinkage and selection operator (LASSO), a shrinkage with selection procedure that provides a more parsimonious approach to covariate selection and model fitting (Houwelingen 2001; Tibshirani 1996). This method minimizes the usual sum of squared errors, with a bound on the sum of the absolute values of the coefficients, thereby shrinking very unstable estimates toward zero, excluding redundant/irrelevant covariates, and avoiding overfitting (Zhao and Yu 2006). We used Sobel's indirect test to assess the influence of child behaviors on the estimates of CPF effect (MacKinnon et al. 2002; Sobel 1982). We used Sobel's indirect test to assess mediation (MacKinnon et al. 2002; Sobel 1982). Interaction terms including CPF and each additional covariate were tested in the models. Effect estimates, 95% confidence intervals (CIs), and p-values were calculated for all analytic procedures. Results were considered significant at $p < 0.05$.

11.2 RESULTS

The retention rate for the full cohort was 82% at the 7-year follow-up, with no significant sociodemographic differences between subjects retained in the study and those lost to follow-up (data not shown). Table 1 lists characteristics of the study sample with complete data on all variables (n = 265). Study families were predominantly low income, with 31% of mothers failing to complete high school by child's age 7 years, and 66% never married. The sample was largely full term (only 4% of children in the sample were < 37 weeks gestational age at delivery) and included very few low-birth-weight infants because a) we excluded high-risk pregnancies from the study cohort, and b) the timing of air monitoring in the third trimester of pregnancy eliminated early deliveries.

CPF exposure levels ranged from nondetectable to 63 pg/g. We imputed exposure levels in participants with nondetectable CPF (n = 115, 43%) according to assay-specific LOD values, with 93 subjects having LOD equal to 0.5 pg/g and 22 subjects having LOD equal to 1 pg/g.

Correlation analyses for exposures and cognitive outcomes. Unadjusted correlations between prenatal CPF exposure and log-transformed WISC-IV Composite Indices (Verbal Comprehension, Working Memory, Processing Speed, and Perceptual Reasoning), and Full-Scale IQ showed significant inverse associations between CPF exposure and a) Working Memory (r = –0.21, p = < 0.0001) and b) Full-Scale IQ (r = –0.13, p = 0.02). We observed a weak inverse correlation between CPF and Perceptual Reasoning (r = –0.09, p = 0.09), while associations of CPF with Verbal Comprehension (r = –0.04) and Processing Speed (r = –0.01) had p-values > 0.05.

TABLE 1: Demographic characteristics of the sample at 7-year follow-up (n = 265).

Characteristic	n (%) or mean ± SD (range)
Home quality[a]	40.23 ± 4.81 (23–52)
Income	
<$20,000	138 (52)
≥$20,000	127 (48)
Maternal education[b]	
Years	12.22 ± 2.58 (1–20)
<High school degree	82 (31)
High school degree	183 (69)
Maternal IQ[c]	85.97 ± 13.46 (60–135)
Maternal race/ethnicity	
Dominican	146 (55)
African American	119 (55)
Marital status	
Never married	175 (66)
Ever married	90 (34)
Child sex	
Male	117 (44)
Female	148 (56)
Gestational age (weeks)	39.3 ± 1.5 (30–43)
Birth weight (g)	3389.8 ± 493.5 (1,295–5,110)
Child age at testing (months)	85.97 ± 2.65 (74.90–101.5)
Prenatal chemical exposures	
Ets[e]	
Exposed	93 (35)
Not exposed	172 (65)
Cotinine (ng/mL)[f]	0.25 ± 0.92 (0.01–8.78)
Lead (µg/dL)[f]	1.09 ± 88 (0.15–7.45)
CPF (pg/g)[f]	3.17 ± 4.671 (0.09–32)
PAHs (ng/m³)[g]	3.37 ± 3.51 (0.50–36.5)

[a]*As measured by the HOME inventory.* [b]*Completed years of education at child's age 7 years.* [c]*As measured by TONI-3.* [d]*Self-reported race/ethnicity (African America = 1; Dominican = 0).* [e]*Self-reported ever exposed to secondhand smoke in pregnancy (yes = 1, no = 2).* [f]*Measured in cord blood.* [g]*Measured by personal air sampling.*

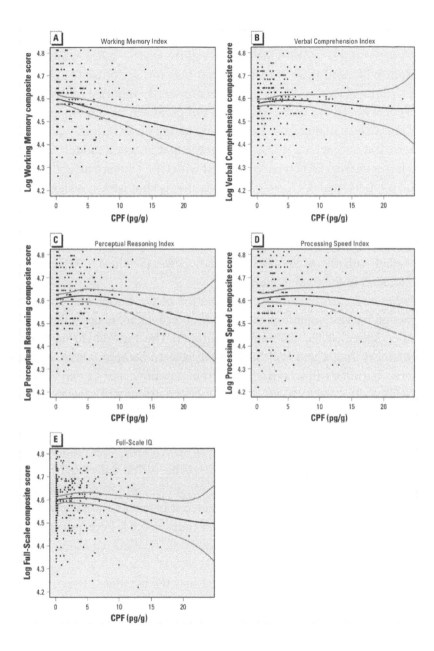

FIGURE 1: Smoothed cubic splines, superimposed over scatterplots, examining the shape of the associations between CPF exposure and (A) Working Memory Index, (B) Verbal Comprehension Index, (C) Perceptual Reasoning Index, (D) Processing Speed Index, and (E) Full-Scale IQ.

Umbilical cord lead was not significantly correlated with CPF level (r = –0.08, p = 0.49) or WISC-IV scores (all p-values > 0.05) among the 89 children with lead data available. Lead was not significantly correlated with CPF level (r = –0.08, p = 0.49, as previously reported by Rauh et al. 2006) or with 7-year WISC-IV scores (all p-values > 0.05) among the 89 children with available data. To avoid excluding observations without lead data, we did not include lead as a covariate in regression models. ETS and (to a lesser extent) PAH were correlated with CPF (Spearman coefficients: 0.113, p = 0.01, and 0.07, p = 0.09, respectively) but were not significantly correlated (using the Mann–Whitney test for the dichotomous ETS variable) with any WISC-IV index (coefficients ranged from –0.02 to 0.03, and p-values ranged from 0.39 to 0.87). Birth weight was not significantly associated with any of the WISC-IV indices (all p-values > 0.05) and was not included in the final models.

Spline regression analysis. Examination of the smoothed cubic spline regression curves, superimposed over scatterplots, indicates subtle differences in shape of the functions (Figure 1). The log-transformed Working Memory Index and Full-Scale IQ appear to be approximately linear, whereas the other functions show some curvature across exposure levels, with sparse observations at the highest exposures. Using the Davidson–MacKinnon test for comparison of non-nested models (Davidson and MacKinnon 1981), we compared models in which CPF was entered as a single continuous outcome with those in which CPF was modeled using B-splines. We failed to reject the null hypothesis that the model with CPF as a continuous measure is adequate against the alternative that the model with CPF modeled using splines provided a better fit for each WISC-IV Index (p-values: Verbal Comprehension Index = 0.07, Perceptual Reasoning Index = 0.08, Processing Speed Index = 0.59, Working Memory Index = 0.40, and Full-Scale IQ = 0.08).

Estimation of linear models. Table 2 lists the estimated B-coefficients, 95% CIs, and p-values for the exposure variable and covariates for the best-fitting linear regression models predicting each WISC-IV outcome. Table 2 also includes the results of linear model selection using the LASSO technique, which eliminates covariates with unstable estimates and results in more parsimonious models. Because the LASSO method uses bootstrapping to obtain standard errors, the coefficient of any covariate may be shrunk to zero if that covariate is an unstable predictor—that is, if its significance depends on the particular subset of data used in the model. The two approaches yielded very similar estimates of CPF effect. Differences in estimates for the covariates in the two methods suggest that the contribution of some covariates to WISC-IV scores may be less stable. Results for both approaches show that, on average, a 1-pg/g increase in CPF is associated with a decrease of –0.006 points in the log-transformed Working Memory score and a decrease of –0.003 points in the log-transformed Full-Scale IQ score. Because of the log transformation, estimated associations between CPF and actual Working Memory and Full-Scale IQ scores vary across the continuum of scores, such that the estimated deficit in the Working Memory score with a 1-pg/g increase in CPF ranges between 0.35 and 0.81 points, and the estimated decrease in Full-Scale IQ is between 0.20 and 0.40 points. The magnitude of these effects is more easily understood by calculating the neurodevelopmental deficit associated with an increase in CPF exposure equal to 1 SD (4.61 pg/g). On average, for each standard deviation increase in exposure, Full-Scale IQ declines by 1.4% and Working Memory declines by 2.8%. We found no significant interactions between CPF and any of the potential or final covariates, including the other chemical exposures measured during the prenatal period (ETS and PAH). Full model results for the linear regressions are provided in Supplemental Material, Table 1 (doi:10.1289/ehp.1003160), for the reader who is interested in the estimates of association between the covariates and outcomes for all of the WISC-IV index scales.

Sensitivity analysis of additional influences on Working Memory Index. To determine whether the observed CPF effect on the Working Memory Index was partially explained by its effect on general intelligence, we added the log-transformed General Ability Index, a general intelligence scale that does not include the Working Memory Index or Processing Speed Index, to the linear regression model. Although the estimate of the General Ability Index effect on Working Memory Index was significant (B-coefficient = 0.57; 95% CI,0.44–0.70; $p < 0.001$), the estimate of the CPF effect remained unchanged (–0.006), and we found no evidence of interaction between CPF and General Ability Index ($p > 0.05$), suggesting that the Working Memory effect is targeted and does not depend upon level of general intelligence.

TABLE 2: Estimated associations between CPF (pg/g) and log-transformed Full-Scale IQ and each of four Composite Index scores from the WISC-IV from LASSO[a] and fully adjusted[b] linear regression models (n = 265).

WISC-IV scale[c]	B-coefficient[c]	95% CI	p-value
Full-scale IQ			
LASSO	−0.003	−0.006 to 0.001	0.064
Fully adjusted	−0.003	−0.006 to 0.000	0.048
Working Memory Index			
LASSO	−0.006	−0.009 to −0.002	<0.001
Fully adjusted	−0.006	−0.010 to −0.002	0.003
Verbal Comprehension Index			
LASSO	NA[d]	NA	NA
Fully adjusted	−0.002	−0.005 to 0.001	0.208
Perceptual Reasoning Index			
LASSO	NA	NA	NA
Fully adjusted	−0.002	−0.006 to −0.002	0.290
Processing Speed Index			
LASSO	NA	NA	NA
Fully adjusted	0.001	−0.004 to −0.005	0.728

NA, not assessed. [a]LASSO models were adjusted for maternal education, maternal IQ, and the HOME Inventory. [b]Fully adjusted models were adjusted for child sex, race/ethnicity, maternal IQ, maternal education income, child age at testing (months), ETS, and PAH.[c]All scales were ln transformed. [d]CPF was not retained in the final LASSO model.

Because child performance on the Working Memory Index can be influenced by child behavior problems (Wechsler 2003), we conducted a supplementary analysis to rule out the possibility that the observed associations between CPF and the Working Memory Index might be affected by behavior problems, as measured by the clinically oriented diagnostic and statistical manual scales on the Child Behavior Checklist. We found no evidence of indirect "mediation" using Sobel's test, with p-values ranging from 0.31 to 0.99 (MacKinnon et al. 2002; Sobel 1982). Full model results are provided in Supplemental Material, Table 2 (doi:10.1289/ehp.1003160), for the reader who is interested in the estimates of association between child behavior problems and Working Memory Index.

Sensitivity analysis of the influence of LOD imputation. After obtaining all results, we recomputed all estimates of association between CPF and WISC-IV scores among subjects with detectable CPF levels only. Analysis with detects alone is known to give unbiased estimates of the parameters of interest (Little 1992). In the present sample, we observed no consistent differences in estimates when we excluded imputed CPF data (data not shown).

11.3 DISCUSSION

Results of this study showed that higher prenatal CPF exposure, as measured in umbilical cord blood plasma, was associated with decreases in cognitive functioning on two different WISC-IV indices, in a sample of urban minority children at 7 years of age. Specifically, for each SD increase in exposure (4.61 pg/g), Full-Scale IQ declined, on average, by 1.4% (0.94–1.8 points) and Working Memory Index scores declined by 2.8% (1.6–3.7 points). The dose–effect relationships between CPF exposure and log-transformed Working Memory Index and Full-Scale IQ scores are linear across the range of exposures in the study population, with no evidence for a threshold. Of the WISC-IV indices used as end points, the Working Memory Index was the most strongly associated with CPF exposure in this population.

Although no other epidemiologic studies have evaluated the neurotoxicity of prenatal CPF exposure on cognitive development at the time of school entry, several prior studies, using the present biomarker of exposure, have reported evidence of early cognitive and behavioral effects associated with a urinary biomarker of nonspecific OP exposure (Engel et al. 2007; Eskenazi et al. 2007; Young et al. 2005). Outcomes associated with exposure in these studies, as well as in our own earlier work (Rauh et al. 2006), have included attentional problems (e.g., Marks et al. 2010). These prior findings are consistent with the present 7-year results, because working memory skills involve attentional processes. More important, problems in working memory may interfere with reading comprehension, learning, and academic achievement, although general intelligence remains in the normal range (Blair 2006). Working memory is less likely than full-scale IQ to be affected by socioeconomic or cultural conditions (Wechsler 2003), providing a useful, more targeted measure of possible neurotoxic effects on brain function.

Several different theories or models address how working memory operates in the human brain, but most agree that it involves a system of limited attention capacity, supplemented by more peripherally based storage systems (Baddeley and Logie 1999). Some theories emphasize the role of attentional control in working memory (e.g., Cowan 1999), whereas others stress a multicomponent model, including a control system of limited attentional capacity (the central executive control system), assisted by phonological and visuospatial storage systems (see review by Baddeley 2003). To date, most studies of the anatomical localization of working memory problems are based on clinical populations (individuals with specific brain lesions) (Vallar and Pagano 2002) and some neuroimaging studies in small numbers of normal subjects (Smith and Jonides 1997). More refined neuropsychological tests and neuroimaging studies are needed to determine whether CPF-related working memory deficits are primarily auditory (part of a phonological loop with implications for language acquisition) or primarily related to visuospatial short-term memory (reflecting nonverbal intelligence tasks).

Few human studies have focused on possible mechanisms underlying neurodevelopmental deficits associated with OP exposure, but there is evidence that certain genetic polymorphisms can affect CPF metabolism (e.g., Berkowitz et al. 2004). Such findings suggest that some populations may be more vulnerable and may exhibit adverse neurodevelopmental effects at much lower exposures than other populations (Berkowitz et al. 2004). Again, neuroimaging studies would be useful to determine if population differences in vulnerability to CPF are also reflected in population differences in brain abnormalities associated with exposure.

Although behavioral alterations observed in rodents may be imperfect analogues for humans, they have guided human studies by identifying specific deficits in locomotor activity, learning, and memory (e.g., Aldridge et al. 2005). In light of experimental evidence suggesting that CPF effects in rodents may be irreversible (Slotkin 2005), it will be important to determine how any neurocognitive deficits associated with prenatal CPF exposure might respond to treatment or early intervention. Here, we may benefit from studies of lead-exposed children that have demonstrated evidence of reversals in learning deficits as a result of environmental enrichment (Guilarte et al. 2003).

Some limitations of this study should be noted. Our sample consists of low-income, urban, minority children who may experience other unmeasured exposures or underlying health problems that could potentially confound or modify associations with pesticide exposure. Furthermore, in the absence of firm mechanistic evidence linking brain anomalies to more refined neuropsychological testing, the observed functional deficits at 7 years of age should be interpreted with caution. We cannot directly compare our findings with the results from the other epidemiological studies that have relied on urinary OP concentrations as the biomarker of exposure.

In June 2000, the U.S. EPA announced a phase-out of the sale of CPF for indoor residential use, with a complete ban effective 31 December 2001 (U.S. EPA 2000, 2002). After the ban, levels of CPF in personal and indoor air samples in our own cohort decreased by more than 65%, and plasma blood levels dropped by more than 80% (Whyatt et al. 2005), despite some lingering residential residues (Whyatt et al. 2007). From other parts of the country, there is evidence of continued low-dose exposures in children from food residues (Lu et al. 2006). Because agricultural use of CPF is still permitted in the United States, it is important that we continue to monitor the levels of exposure in potentially vulnerable populations, including pregnant women in agricultural communities, and evaluate the long-term neurodevelopmental implications of exposure to CPF and other OP insecticides.

REFERENCES

1. Achenbach TM, Rescorla LA. 2001. Manual for the ASEBA School-Age Forms and Profiles. Burlington, VT:University of Vermont, Research Center for Children, Youth and Families.
2. Aldridge JE, Meyer A, Seidler FJ, Slotkin TA. 2005. Alterations in central nervous system serotonergic and dopaminergic synaptic activity in adulthood after prenatal or neonatal chlorpyrifos exposure. Environ Health Perspect 113:1027–1031.
3. Baddeley A.. 2003. Working memory: looking back and forward. Nat Rev Neurosci 4(10):829–839.

4.   Baddeley AD, Logie RH. 1999. Working memory: the multi-component model. In: Models of Working Memory: Mechanisms of Active Maintenance AND Executive Control (Miyake A, Shah P, eds) New York:Cambridge University press, 28–61.

5.   Barr DB, Barr JR, Maggio VL, Whitehead RD Jr, Sadowski MA, Whyatt RM, et al. 2002. A multi-analyte method for the quantification of contemporary pesticides in human serum and plasma using high resolution mass spectrometry. J Chromatog B Anal Technol Biomed Life Sciences 778:99–111.

6.   Berkowitz GS, Obel J, Deych E, Lapinski R, Godbold J, Liu Z, et al. 2003. Exposure to indoor pesticides during pregnancy in a multiethnic, urban cohort. Environ Health Perspect 111:79–84.

7.   Berkowitz GC, Wetmur JG, Birman-Deych E, Obel J, Lapinski RH, Godbold JH, et al. 2004. In utero pesticide exposure, maternal paraoxinase activity, and head circumference. Environ Health Perspect 112:388–391.

8.   Blair C.. 2006. How similar are fluid cognition and general intelligence? A developmental neuroscience perspective on fluid cognition as an aspect of human cognitive ability. Behav Brain Science 29:109–125.

9.   Bradman A, Barr DB, Claus Henn BG, Drumheller T, Curry C, Eskenazi B. 2003. Measurement of pesticides and other toxicants in amnionic fluid as a potential biomarker of prenatal exposure: a validation study. Environ Health Perspect 113:1802–1807.

10.   Brown L, Sherbenou RJ, Johnson SK. 1997. Test of Nonverbal Intelligence: A Language-Free Measure of Cognitive Ability. 3rd ed. Austin, TX:PRO-ED, Inc.

11.   Caldwell BM, Bradley RH. 1979.

12.   Chiodo LM, Jacobson SW, Jacobson JL. 2004. Neurodevelopmental effects of postnatal lead exposure at very low levels. Neurotoxicol Teratol 26:359–371.

13.   Cowan N. 1999. An embedded-processes model of working memory. In: Models of Working Memory: Mechanisms of Active Maintenance and Executive Control (Miyake A, Shah P, eds). New York:Cambridge University Press, 62–101.

14.   Davidson R, MacKinnon J.. 1981. Several tests for model specification in the presence of alternative hypotheses. Econometrica 49:781–793.

15.   Engel SM, Berkowitz GS, Barr DB, Teitelbaum SL, Siskind J, Meisel SL, et al. 2007. Prenatal organophosphate metabolite and organochlorine levels and performance on the Brazelton Neonatal Behavioral Assessment Scale in a multiethnic pregnancy cohort. Am J Epidemiol 165:1397–1404.

16.   Eskenazi B, Marks AR, Bradman A, Harley K, Barr DB, Johnson C, et al. 2007. Organophosphate pesticide exposure and neurodevelopment in young Mexican-American children. Environ Health Perspect 115:792–798.

17.   Guilarte TR, Toscano CD, McGlothan JL, Weaver SA. 2003. Environmental enrichment reverses cognitive and molecular deficits induced by developmental lead exposure. Ann Neurol 53:50–56.

18.   Hecht SA, Burgess SR, Torgesen RK, Wagner RK, Rashotte CA. 2000. Explaining social class differences in growth of reading skills from beginning kindergarten through fourth-grade: the role of phonological awareness, rate of access and print knowledge. Read Writ Interdisciplin J 12:99–127.

19.   Houwelingen JC. 2001. Shrinkage and penalized likelihood as methods to improve predictive accuracy. Stat Neerl 55(1):17–34.

20. Jusko TA, Henderson CR, Lanphear BP, Cory-Slechta DA, Parsons PJ, Canfield RL. 2008. Blood lead concentrations < 10 μg/dL and child intelligence at 6 years of age. Environ Health Perspect 116:243–248.

21. Landrigan PJ, Schechter CB, Lipton JM, Fahs MC, Schwartz J. 2002. Environmental pollutants and disease in American children: estimates of morbidity, mortality, and costs for lead poisoning, asthma, cancer, and developmental disabilities. Environ Health Perspect 110:721–728.

22. Levin ED, Addy N, Baruah A, Elias A, Cannelle CN, Seidler FJ, et al. 2002. Prenatal chlorpyrifos exposure in rats causes persistent behavioral alternations. Neurotoxicol Teratol 24:733–741.

23. Little RJA. 1992. Regression with missing X's: a review. J Am Stat Assoc 87:1227–1237.

24. Lu C, Toepel K, Irish R, Fenske RA, Barr DB, Bravo R. 2006. Organic diets significantly lower children's dietary exposure to organophosphate pesticides. Environ Health Perspect 114:260–263.

25. MacKinnon DP, Lockwood CM, Hoffman JM, West SG, Sheets V. 2002. A comparison of methods to test mediation and other intervening variable effects. Psychol Methods 7:83–104.

26. Marks AR, Harley K, Bradman A, Kogut K, Barr DB, Johnson C, et al. 2010. Organophosphate pesticide exposure and attention in young Mexican-American children: the CHAMACOS Study. Environ Health Perspect 118:1768–1774.

27. Perera FP, Illman SM, Kinney PL, Whyatt RM, Kelvin EA, Shepard P, et al. 2002. The challenge of preventing environmentally related disease in young children: community-based research in New York City. Environ Health Perspect 110:197–204.

28. Perera FP, Rauh V, Tsai WY, Kinney P, Camann D, Barr D, et al. 2003. Effects of transplacental exposure to environmental pollutants on birth outcomes in a multiethnic population. Environ Health Perspect 111:201–205.

29. Qiao D, Seidler FJ, Tate CA, Cousins MM, Slotkin TA. 2003. Fetal chlorpyrifos exposure: adverse effects on brain cell development and cholinergic biomarkers emerge postnatally and continue into adolescence and adulthood. Environ Health Perspect 111:536–544.

30. R Development Core Team 2009. R: A Language and Environment for Statistical Computing. Vienna:R Foundation for Statistical Computing.

31. Rauh VA, Garfinkel R, Perera FP, Andrews H, Barr D, Whitehead D, et al. 2006.

32. Rauh VA, Whyatt RM, Garfinkel R, Andrews H, Hoepner L, Reyes A, et al. 2004. Developmental effects of exposure to environmental tobacco smoke and material hardship among inner-city children. Neurotoxicol Teratol 26:373–385.

33. Richardson DB, Ciampi A. 2003. Effects of exposure measurement error when an exposure variable is constrained by a lower limit. Am J Epidemiol 57(4):355–363.

34. Richardson RJ. 1995. Assessment of the neurotoxic potential of chlorpyrifos relative to other organophosphorus compounds: a critical review of the literature. J Toxicol Environ Health 44:135–165.

35. Rothenberg SJ, Rothenberg JC. 2005. Testing the dose-response specification in epidemiology: public health and policy consequences for lead. Environ Health Perspect 113:1190–1195.

36. Slotkin TA. 2004. Cholinergic systems in brain development and disruption by neurotoxicants: nicotine, environmental tobacco smoke, organophosphates. Toxicol Appl Pharmacol 198:132–151.

37. Slotkin TA. 2005. Developmental neurotoxicity of organophosphates: a case study of chlorpyrifos. In: Toxicity of Organophosphates and Carbamate Pesticides (Gupta RC, ed). San Diego:Academic Press, 293–314.

38. Slotkin TA, Seidler FJ. 2005. The alterations in CNS serotonergic mechanisms caused by neonatal chlorpyrifos exposure are permanent. Dev Brain Res 158:115–119.

39. Smith EE, Jonides J. 1997. Working memory: a view from neuroimaging. Cogn Psychol 33:5–42.

40. Sobel ME. 1982. Asymptotic confidence intervals for indirect effects in structural equation models. Sociol Methodol 13:290–312.

41. Tibshirani R.. 1996. Regression shrinkage and selection via the lasso. J R Stat Soc Series B Stat Methodol 58(1):267–288.

42. U.S. EPA (U.S. Environmental Protection Agency) 2000. Chlorpyrifos Revised Risk Assessment and Agreement with Registrants. Washington, DC:U.S. EPA.

43. U.S. EPA (U.S. Environmental Protection Agency) 2002. Chlorpyrifos End-Use Products Cancellation Order. Washington, DC:U.S. EPA.

44. Vallar G, Pagano C. 2002. Neuropsychological impairments of verbal short-term memory. In: Handbook of Memory Disorders (Baddeley AD, Kopelman MD, Wilson BA, eds). Chichester, UK:Wiley, 249–270.

45. Wechsler D. 2003.

46. Whyatt RM, Barr DB, Camann DE, Kinney PL, Barr JR, Andrews HF, et al. 2003. Contemporary-use pesticides in personal air samples during pregnancy and blood samples at delivery among urban minority mothers and newborns. Environ Health Perspect 111:749–756.

47. Whyatt RM, Camann D, Perera FP, Rauh VA, Tang D, Kinney PL, et al. 2005. Biomarkers in assessing residential insecticide exposures during pregnancy and effects on fetal growth. Toxicol Appl Pharmacol 206:246–254.

48. Whyatt RM, Camann DE, Kinney PL, Reyes A, Ramirez J, Dietrich J, et al. 2002. Residential pesticide use during pregnancy among a cohort of urban minority women, Environ Health Perspect 110:507–514.

49. Whyatt RM, Garfinkel R, Hoepner LA, Holmes D, Borjas M, Perera FP, et al. 2007. Within and between home variability in indoor-air insecticide levels during pregnancy among an inner-city cohort from New York City, Environ Health Perspect 115:383–390.

50. Whyatt RM, Rauh VA, Barr DB, Camann DE, Andrews HF, Garfinkel R, et al. 2004. Prenatal insecticide exposure and birth weight and length among an urban minority cohort. Environ Health Perspect 112:1125–1132.

51. Young JG, Eskenazi B, Gladstone EA, Bradman A, Pedersen L, Johnson C, et al. 2005. Association between in utero organophosphate exposure and abnormal reflexes in neonates. Neurotoxicology 26:199–209.

52. Zhao P, Yu B.. 2006. On model selection consistency of Lasso. J Machine Learn Res 7:2541–2563.

There are several supplemental files that are not available in this version of the article. To view this additional information, please use the citation on the first page of this chapter.

PART VI

FLAME RETARDANTS

LITERATURE

CHAPTER 12

PRENATAL EXPOSURE TO PBDES AND NEURODEVELOPMENT

JULIE B. HERBSTMAN, ANDREAS SJLIDIN,
MATTHEW KURZON, SALLY A. LEDERMAN, RICHARD S. JONES,
VIRGINIA RAUH, LARRY L. NEEDHAM, DELIANG TANG,
MEGAN NIEDZWIECKI, RICHARD Y. WANG,
AND FREDERICA PERERA

Polybrominated diphenyl ethers (PBDEs) are widely used flame retardant compounds applied to a wide array of textiles, building materials, and electronic equipment, including computers and televisions. Because they are additives rather than chemically bound to consumer products, they have the propensity to be released into the environment (Darnerud et al. 2001). PBDEs are persistent organic chemicals, and some congeners can bioaccumulate; therefore they have become ubiquitous contaminants detectable in the environment, in animals, and in humans (Hites 2004; Sjodin et al. 2008b).

A number of toxicologic studies have demonstrated that exposure to PBDEs may have endocrine-disrupting effects. Most of these studies have focused on thyroid hormone disruption and a smaller number on disruption of the estrogen/androgen hormone system [reviewed by Darnerud

Used with permission from Environmental Health Perspectives. Herbstman JB, Sjödin A, Kurzon M, Lederman SA, Jones RS, Rauh V, Needham LL, Tang D, Niedzwiecki M, Wang RY, Perera F. Prenatal Exposure to PBDEs and Neurodevelopment. Environmental Health Perspectives, **118** *(2010), http:// dx.doi.org/10.1289/ehp.0901340.*

(2008)]. Endocrine disruption during critical developmental periods may result in irreversible effects on differentiating tissue, including the brain (Bigsby et al. 1999). Causal relationships between prenatal exposure to PBDEs and indices of developmental neurotoxicity have been observed in experimental animal models [reviewed by Costa and Giordano (2007)]. Thus, the disruption of endocrine pathways by prenatal exposure to hormonally active environmental chemicals may affect neurodevelopment in children.

Although the association between prenatal exposure to PBDEs and adverse neurodevelopmental effects has been observed in animal models, it has not been adequately explored in human populations. In a longitudinal cohort study initiated by the Columbia Center for Children's Environmental Health (CCCEH), we examined the impact of prenatal exposures to selected toxicants, including PBDEs, that may be present in the ambient environment but may also have been emitted from the World Trade Center (WTC) buildings in New York City after the 11 September 2001 (9/11) terrorist attack. Here we report the relationship between prenatal PBDE and polybrominated biphenyl (PBB-153) measured in umbilical cord blood in humans and indicators of neurodevelopment at 12–48 and 72 months of age.

12.1 METHODS

Study population. We established a prospective cohort study of women who were pregnant on 11 September 2001 and subsequently delivered at one of three downtown hospitals including Beth Israel, St. Vincent's (and St. Vincent's affiliated Elizabeth Seton Childbearing Center), which are all approximately 2 miles from the WTC site, and New York University Downtown Hospital, which is within a half-mile of the WTC site. The study methods have been described previously (Lederman et al. 2004). In brief, beginning 12 December 2001 [when institutional review board (IRB) approval was obtained], women were approached in the hospital when they presented for labor and delivery. The women were briefly screened for eligibility, recruited, and enrolled, and they consented before delivery. This study was conducted in accordance with all applicable requirements of the United States (including IRB approval), and all human participants gave written informed consent before participation in this study. Eligible women included those who were between 18 and 39 years of age, reported smoking < 1 cigarette per day during pregnancy, were pregnant on 11 September 2001 (based on their estimated date of conception), and reported no diabetes, hypertension, HIV infection or AIDS, or use of illegal drugs in the preceding year. Not all mothers agreed to have their child followed after birth. For example, some of the Chinese children were to be raised in China [see Supplemental Material, Table 1 (doi:10.1289/ehp.0901340) for follow-up information].

TABLE 1: Concentrations (ng/g lipid) of PBDEs and BB-153 in cord blood.

	Cord blood measurements				Cord measurements with >1 neurodevelopmental test (n = 152)			
	n	% > LOD	Median	Maximum	n	% > LOD	Median	Maximum
BDE-47	219	81.4	11.2	613.1	152	83.6	11.2	613.1
BDE-85	189	18.5	0.7	16.6	141	17.7	0.7	16.6
BDE-99	210	59.5	3.2	202.8	152	57.9	3.2	202.8
BDE-100	209	63.6	1.4	71.9	152	69.1	1.4	71.9
BDE-153	201	49.8	0.7	28.9	143	55.9	0.7	28.9
BDE-154	200	6.0	0.6	11.1	146	6.2	0.6	11.1
BDE-183	204	3.9	0.6	2.8	147	4.1	0.6	2.8
BB-153	197	11.2	0.6	8.0	145	13.1	0.9	8.0

Data collection. Medical records of the mother and newborn were abstracted for information relating to pregnancy, delivery, and birth outcomes. Interviews were conducted (generally the day after delivery) by bilingual interviewers in the preferred or native language (English, Spanish, or Chinese) of the participants. Demographic information, reproductive history, background environmental exposures, occupational history, and the location of the residences and workplaces of the woman during each of the 4 weeks after 11 September 2001 were determined during this interview. Maternal intelligence was measured using the Test of Non-Verbal Intelligence, Second Edition (TONI-2), a 15-min, language-free measure of general intelligence that is relatively stable and free of cultural bias (Brown et al. 1990).

Developmental assessment. When the children were approximately 12, 24, and 36 months of age, the Bayley Scales of Infant Development, Second Edition (BSID-II) were administered, providing scores from the Mental Development Index (MDI) and the Psychomotor Development Index (PDI). The BSID-II is a widely used developmental test designed for children 12–42 months of age that is norm-referenced and can be used to identify children with developmental delay. The assessment provides a developmental quotient (raw score/chronological age), generating a continuous MDI or PDI score, both with mean ± SD = 100 ± 15.

When the children were 48 and 72 months of age, the Wechsler Preschool and Primary Scale of Intelligence, Revised Edition (WPPSI-R) was administered, which measures cognitive development and contains verbal and nonverbal performance tests. We used the WPPSI-R because of its availability in Chinese, rather than the third edition.

Not all children were available for all developmental assessments, resulting in different numbers of children tested at each age. Assessments were conducted in the first language of the child (English or Chinese) by trained research technicians. In some cases, when the primary language of the child was not English or Chinese (e.g., Yiddish), we relied on maternal translation. Statistical analyses for this study were conducted with and without the children for whom the child's primary language was not English or Chinese (n = 30), and the results were similar (data not shown).

Most assessments were conducted at the CCCEH. However, a proportion of the assessments were conducted in the child's home if the parents were unable or unwilling to come to the center to complete the follow-up [see Supplemental Material, Table 1 (doi:10.1289/ehp.0901340) for details].

Blood collection. Umbilical cord blood was collected at delivery, and maternal blood was typically collected on the day after delivery. On average, 30.7 mL blood was collected from the umbilical cord, and 30–35 mL blood was collected from the mothers. Samples were transported to the laboratory and processed within several hours of collection. The buffy coat, packed red blood cells, and plasma were separated and stored at −70°C. Frozen plasma from 210 cord samples was transferred on dry ice to the Centers for Disease Control and Prevention for laboratory analyses for the PBDEs and PBB-153. The concentrations of these chemicals in the cord blood were used as an indicator of fetal exposure during gestation (Mazdai et al. 2003; Qiu et al. 2009).

Laboratory methods. Details regarding the analysis of the plasma samples for PBDEs are given elsewhere (Hovander et al. 2000; Sjodin et al. 2004). Briefly, the samples were automatically fortified with ^{13}C-labeled internal standards. The samples were subjected to an initial liquid/liquid extraction with hexane:methyl-tert-butyl ether after denaturation with 1 M HCl and isopropanol (Hovander et al. 2000). Thereafter, coextracted lipids were removed on a silica:silica/sulfuric acid column using the Rapid Trace equipment (Zymark, Hopkinton, MA) for automation. Final determination of the target analytes was performed by gas chromatography-isotope dilution high-resolution mass spectrometry employing an MAT95XP (ThermoFinnigan MAT, Bremen, Germany) instrument (Sjodin et al. 2004). Concentrations of target analytes were reported as picograms per gram whole weight (weight of plasma) and nanograms per gram lipid weight (weight of plasma lipids). The plasma lipid concentrations were determined using commercially available test kits from Roche Diagnostics Corp. (Indianapolis, IN) for the quantitative determination of total triglycerides (product no. 011002803-0600) and total cholesterol (product no. 011573303-0600). Final determinations were made on a Hitachi 912 Chemistry Analyzer (Hitachi, Tokyo, Japan). Limits of quantification were determined in relation to the method blanks and in relation to the quantification limit of the instrument, which is proportional to the sample size. Cotinine concentrations were measured in cord and maternal blood by use of liquid chromatography in conjunction with atmospheric pressure ionization tandem mass spectrometry (Bernert et al. 1997).

The plasma samples were analyzed for the following PBDE congeners (by International Union of Pure and Applied Chemistry numbers): 2,2,2′,4,4′-tetraBDE (BDE-47); 2,2′,3,4,4′-pentaBDE (BDE-85); 2,2′,4,4′,5-pentaBDE (BDE-99); 2,2′,4,4′,6-pentaBDE (BDE-100); 2,2′,4,4′,5,5′-hexaBDE (BDE-153); 2,2′,4,4′,5,6′-hexaBDE (BDE-154); 2,2′,3,4,4′,5′,6-heptaBDE (BDE-183); and 2,2′,4,4′,5,5′-hexaBB (BB-153).

Quality control/quality assurance. We determined background levels by measuring the level of target analytes in blank samples in the same run as the study samples (three blanks per 24 study samples). All concentrations reported were corrected for the average amount present in the blank samples. The limit of detection (LOD) when no analytical background was detected in blank samples was defined as a signal-to-noise ratio > 3. When an analytical background was detected in the blanks, the LOD was defined as three times the SD of the blanks.

The plasma samples used in this cohort were not collected solely for the purpose of PBDE analysis. Therefore, we examined the ratio of BDE-99 over BDE-47 for any indication of contamination from indoor particulate matter, with the assumption that a high ratio would indicate sample contamination during sample collection. The median ratio of BDE-99 over BDE-47 is 1.2 in residential dust samples (Sjodin et al. 2008a), whereas in human samples this ratio is typically significantly lower. In the 2003–2004 National Health and Nutrition Examination Survey (NHANES), the median ratio of BDE-99 to BDE-47 was 0.23, whereas the 95th percentile of this ratio was 0.43 (Sjodin et al. 2008b). In our study, we found that 16 of 210 samples had a BDE-99 to BDE-47 ratio > 0.43, corresponding to 7.6% of the samples. This frequency of samples having a ratio > 0.43 is similar to that of the NHANES survey, and we can thus conclude that no detectable contamination occurred during the collection of the cord samples in this study. Statistical analyses for this study were conducted both including and excluding the aberrant samples (n = 16), and the results were similar (data not shown).

WTC exposure. In previous analyses in this cohort, we used two indices to describe exposure to the WTC: geographic proximity to the WTC during the first month after 9/11 and timing of exposure relative to date of delivery. We found that women who lived closest to the WTC during the first month after 9/11 (constituting the group we would estimate to have the largest exposure to the WTC) did not have higher concentrations of PBDEs compared with those who lived farther from the towers in the first 4 weeks after the attack. However, we found that women who delivered sooner after 9/11 (constituting the group who were further along in their pregnancy on 9/11) tended to have higher cord blood concentrations of PBDEs (unpublished data). Because we quantify prenatal PBDE exposure using a biological marker that integrates exposure from all sources, the source of the PBDEs is not relevant to the effect of prenatal exposure to PBDEs on neurodevelopment. To our knowledge, PBDEs are not associated with any other neurotoxic exposure that could confound the observed associations.

TABLE 2: Characteristics of all cohort members (n = 329), participants with cord blood measurement of PBDEs (n = 210), and those included in our study sample (n = 152).

	All participants (n = 329)	Cord PBDEs (n = 210)	Cord measurements >1 neurodevelopmental tests (n = 152)
Maternal characteristics			
Maternal age (years)	30.2 ± 5.2	30.4 ± 5.1	31.2 ± 4.9**
Maternal education			
< High school	61 (18.5)	45 (21.4)	21 (13.8)
High school	56 (17.0)	36 (17.1)	25 (16.4)
Some college	73 (22.2)	46 (21.9)	34 (22.4)
Four year college degree	72 (21.9)	41 (19.5)	34 (22.4)
Post college education	67 (20.4)	42 (20.0)	38 (25.0)
Race/ethnicity			
Chinese	92 (28.0)	72 (34.3)*	41 (27.0)
Asian (non-Chinese)	21 (6.4)	13 (6.2)	9 (5.9)
Black	50 (15.2)	27 (12.8)	23 (15.1)
White	133 (40.4)	77 (36.7)	62 (40.8)
Other	33 (10.0)	21 (10.0)	17 (11.2)
Married/living with partner	265 (80.6)	172 (81.9)	126 (82.9)
TONI-2 score	95.8 ± 11.4	95.8 ± 11.3	95.8 ± 13.0
Missing TONI	118 (35.9)	82 (39.0)	26 (17.1)**
Maternal exposure to ETS, reported as smoker in the home (%)	59 (17.9)	36 (17.1)	26 (17.1)
Ate fish during the pregnancy	233 (70.8)	150 (71.4)	110 (72.4)
Material hardship	31 (9.4)	20 (9.5)	16 (10.5)
Infant characteristics			
Birth weight (g)	3419.5 ± 469.1	3399.2 ± 472.5	3412.0 ± 487.4
Birth length (cm)	50.8 ± 2.8	50.5 ± 2.7*	50.6 ± 2.7
Birth head circumference (cm)	34.2 ± 1.5	34.2 ± 1.4	34.3 ± 1.5
Gestational age (days)	276.8 ± 9.9	276.4 ± 10.4	276.6 ± 9.5
Male	161 (48.9)	105 (50.0)	77 (50.7)
Proportion of first-year breast-fed (% of 1 year)	0.24 ± 0.28	0.22 ± 0.27	0.26 ± 0.28

TABLE 2: *Cont.*

	All participants (n = 329)	Cord PBDEs (n = 210)	Cord measurements >1 neurodevelopmental tests (n = 152)
Residential characteristics			
Worked and/or lived within 1 mile of the WTC during any of the 4 weeks after 9/11	62 (18.8)	43 (20.5)	32 (21.0)
Worked and/or lived within 2 miles of the WTC during any of the 4 weeks after 9/11	141 (42.8)	94 (44.8)	73 (48.0)

*Values are mean ± SD or n(%). * Statistical comparison between those in the full cohort and those with cord blood measurements , p<0.05. **Statistical comparison between those in the full cohort and those in our study sample, P<0.05*

Statistical methods. Concentrations of PBDEs were lipid- and natural log-adjusted. PBDEs commonly detected in cord blood (detected in > 55% of samples) were handled as continuous variables in the statistical models. This was the case for BDEs 47, 99, 100, and 153. We used the LOD divided by the square root of 2 for concentrations below the LOD. Based on the log-normal distribution of each of these BDE congeners, we also compared participants having cord concentrations in the highest 20% with those in the lowest 80% of the population distribution to evaluate the impact of having exposures at the high end of the exposure distribution. This categorization was selected because it distinguished those with exposures in the tail of the log-normal distribution. The majority of cord samples (> 50%) had levels below the limits of detection for BDEs 85, 154, and 183, and BB-153. We evaluated BDE-85 and BB-153 as dichotomous measures (detected vs. nondetected); we did not analyze BDEs 154 and 183 because only 6% and 4% of the samples, respectively, had detectable concentrations (Table 1).

We generated descriptive statistics and evaluated bivariate associations using analysis of variance and chi-square tests to compare stratum-specific means and proportions, respectively. We examined the data using lowess curves and determined that linear models using natural logarithmic (ln)-transformed PBDE concentrations fit the data well. Therefore, we conducted multivariate linear regression analyses to evaluate the relationships between prenatal PBDE concentrations (using continuous measures for ln-transformed BDEs 47, 99, 100, and 153 and dichotomous measures for BDEs 85, 154, and 183) and continuous scores on developmental tests (MDI and PDI at 12, 24, and 36 months and Full, Verbal, and Performance scores at 48 and 72 months). We were not able to consider developmental test scores as dichotomous measures, using the test-specific recommended cutoffs for defining children as "delayed" or "borderline delayed" because of the small sample size and small number of children who met these criteria.

We selected covariates for inclusion in multivariate models based on their a priori association with neurodevelopment (Tong and Lu 2001), including age at testing, sex of child, ethnicity (Asian, black, white, or other), environmental tobacco smoke (ETS) exposure in the home [yes/no; based on self-report and validated in this data set by cotinine measured in cord blood, using methodology described by Jedrychowski et al. (2009)] and IQ of the mother. We also considered the inclusion of additional covariates if they changed the beta coefficient for PBDEs > 10% when they were added to the a priori set one at a time. This resulted in a final covariate set that added to our a priori set gestational age at birth (based on the best obstetric estimate), maternal age, maternal education, material hardship during pregnancy (defined as having gone without either food, shelter, gas/electric, clothing, or medication/medical care because of financial constraint), and breast-feeding [considering both breast-feeding duration and exclusiveness, defined by Lederman et al. (2008)]. The results of these models are presented as model 1. We also created a model 2, which included all the previous covariates plus two study-specific variables: the language (including whether the mother aided in translation) and location (home or study site) of the interview and assessment. We explored the effects of whether the mother ate fish/seafood when she was pregnant (yes/no) and also the effects of cord blood total mercury and lead concentrations (continuous measures), but found that these covariates did not materially change the relationships between PBDE concentrations and developmental indicators.

To determine whether only a few cases could have a substantial association with the adjusted PBDE regression coefficients, we used AV plots to examine the residuals from the regression lines for the adjusted PBDE regression coefficients to find possible influential cases. These are cases that are outliers both for the outcomes (PDI, MDI, WPPSI) and the independent variables of interest (PBDE compounds). When the possible influential cases were removed, the largest changes in the regression coefficients were < 1 point, with no changes in significance levels, using p < 0.05 as a cut point.

12.2 RESULTS

Median cord concentrations of PBDE congeners 47, 99, and 100 in the full cohort were 11.2, 3.2, and 1.4 ng/g lipid; 81.4%, 59.5%, and 63.6%, respectively, were above the LOD (Table 1). Overall concentrations and the proportion of participants with PBDE concentrations above the LOD were not significantly different in the study subsample. The proportion of participants with detectable concentrations of BDEs 85, 153, 154, and 183 ranged from approximately 4% to 50%. PBDE congeners 47, 99, and 100 were highly intercorrelated (r = 0.74–0.88).

Characteristics of the full cohort (n = 329), the subset of 210 participants with cord PBDE (and PBB) measurements, and the subset of 152 with both cord measurement and a neurodevelopmental test are shown in Table 2. Those with cord blood measurements were similar to the full cohort except that there were proportionally more Chinese participants with cord blood measurements (34.3% compared with 28.0% in the full cohort) and those with measurements were slightly smaller in birth length. The study sample and the full cohort were similar except that mothers included in the study sample were slightly older at the time of delivery (31.2 vs. 30.2 years, p < 0.01) and were more educated (not statistically significant). Those in the study sample were more likely to have completed maternal IQ measurements, which is expected, considering that this measurement was collected at follow-up visits, not at the delivery hospital. There were no differences in the proportions working and/or living closest to the WTC (within 1 or 2 miles) at the time of the attack.

There were 118, 117, 114, 104, and 96 children with available cord PBDE measurements who also had a developmental assessment at 12, 24, 36, 48, and 72 months of age, respectively. For those assessed at all of these time points, the median cord plasma concentrations of BDE-47 and 99 were 12.1 ng/g lipid and 3.5 ng/g lipid. For BDE-100, median concentrations were approximately 1.5 ng/g lipid for children assessed at 12–48 months; for those assessed at 72 months, the median cord blood concentration was 1.4 ng/lipid.

In cross-sectional analyses using multivariate linear regression, prenatal exposure to BDE-47 was negatively associated with neurodevelopmental indices (Table 3). These relationships were statistically significant for 12-month PDI (borderline), 24-month MDI, and 48-month Full and Verbal IQ scores. For every ln-unit change in BDE-47, scores were, on average, 2.1–3.1 points lower on developmental indices. For BDE-99 (Table 4), statistically significant negative associations were detected for 24-month MDI [β = –2.82; 95% confidence interval (CI), –4.86 to –0.78]. Prenatal exposure to BDE-100 was negatively associated with neurodevelopmental indices (Table 5), with statistically significant relationships observed for 24-month MDI, 48-month Full, Verbal, and Performance IQ scores, and 72-month Performance IQ scores. For every ln-unit change in BDE-100, scores were, on average, 3.4–4.0 points lower on developmental indices. For BDE-153 (Table 6), statistically significant negative associations were detected for 48-month and 72-month Full and Performance IQ scores. For every ln-unit change in BDE-153, scores were, on average, 3.1–4.2 points lower. The strength of association between BDE-153 and IQ scores was much larger in the adjusted models compared with the univariate model. It appears that the strong positive association of maternal education between IQs at ages 48 and 72 months was responsible for much of this change.

We also evaluated the difference in mean developmental score comparing children who were in the highest 20% of the prenatal exposure distribution with those in the lower 80% of the distribution for BDEs 47, 99, and 100 (Figure 1). We found that, on average, children with the higher prenatal concentrations of BDEs 47, 99, and 100 scored lower than the rest of the population on nearly all neurodevelopmental indices at all time points (12–48 and 72 months). These differences ranged in magnitude; the largest differences were observed with all three congeners for the 24-month MDI (statistically significant differences of –7.7, –9.3, and –10.9 points for BDEs 47, 99, and 100, respectively) and for 48-month Verbal and Full IQ scores (ranging from –5.5 to –8.0 points). For BDE-

153, adjusting for the same covariate set, those in the highest 20% of the exposure distribution scored, on average, 6.3 points lower at 48 months (95% CI, −13.0 to 0.4) and 8.1 points lower at 72 months (95% CI, −15.6 to −0.6) on the performance IQ scale.

We used multivariate linear regression models to evaluate whether having detectable prenatal concentrations of BDE-85 and BB-153 was significantly related to developmental indices. Adjusting for the exact age of the child at test administration, ethnicity, IQ of the mother, sex of the child, gestational age at birth, maternal age, ETS (yes/no), maternal education, material hardship, and breast-feeding, we found that those with detectable cord concentrations of BDE-85 scored, on average, 11 points lower on the 24-month MDI (95% CI, −17.0 to −5.2); 6.4 points lower on 24-month PDI (95% CI, −11.8 to −0.8); 7.7 points lower on 36-month PDI (95% CI, −15.0 to −0.4); 6.5 points lower on 48-month Verbal IQ (95% CI, −13.3 to 0.2); and 6.9 points lower on 48-month Full IQ (95% CI, −12.8 to −0.9). There were no statistically significant associations between prenatal BB-153 levels with developmental indices measured at any other ages.

12.3 DISCUSSION

We found evidence suggesting that children who had higher cord blood concentrations of BDEs 47, 99, and 100 scored lower on tests of mental and physical development at ages 12–48 and 72 months. These associations were significant for 12-month PDI (BDE-47); 24-month MDI (BDEs 47, 99, and 100); 48-month Full IQ (BDEs 47, 100, and 153); Verbal IQ (BDEs 47 and 100) and Performance IQ (BDEs 100 and 153); and 72-month full and Performance IQ (BDEs 100 and 153). Children who were in the highest 20% of cord blood concentrations of BDEs 47, 99, or 100 had significantly lower developmental scores compared with children who were in the lower 80% of the exposure distributions for these chemicals. These differences were particularly evident at 48 months of age.

Adverse neurodevelopmental effects associated with prenatal PBDE exposure can be detected both at early ages (12–36 months) and as the children age (48 and 72 months). Neurodevelopmental deficits documented by the WPPSI during the preschool period are an important predictor of subsequent academic performance (Kaplan 1993). Documenting the first appearance of potentially longer-term adverse effects at early ages is also important, because these indicators may identify children who could benefit from early intervention programs. The identification of later deficits may indicate the persistence of early effects and/or an increase in the magnitude of effect with age, as has been shown in some animal studies (e.g., Viberg et al. 2003a).

Our results are consistent across congeners and over time. This may be predictable because the PBDE congeners are highly correlated, and for individuals, repeated developmental scores are also correlated. Although the number of participants lost to follow-up between 12 and 72 months was relatively low (81% of subjects available for analyses at 12 months were also assessed at 72 months) and losses are independent of exposure, our overall sample is relatively small. Therefore, even small losses to follow-up may limit our power to detect significant differences in multivariate models. The resulting small sample size precluded the analysis of exposure effects on developmental delay, and we were unable to look at interactions. However, the developmental deficits of the magnitude we observed in this study are likely to have the largest functional impact on those who score at the lower end of the population distribution.

The only other epidemiologic study reporting the neurodevelopmental effects of prenatal exposure to PBDEs was published recently (Roze et al. 2009). In this study of 62 Dutch children, the authors present correlations between exposure to PBDEs (measured during the 35th week of pregnancy) and > 20 indices of child development and behavior at age 60–72 months. The authors report that prenatal PBDE exposure was associated with some adverse effects on development (reduced fine manipulative abilities and increased attentional deficits) as well as some beneficial effects (better coordination, better visual perception, and better behavior). The authors evaluated, but did not find statistically significant, correlations between prenatal PBDE exposures and any of the WPPSI-R domains. Our results are not consistent with these findings. However, important differences in exposure (median exposure in our population was 4 times higher for BDEs 47 and 99; 2.3 times higher for BDE-100; and one-fifth their concentration for BDE-153), sample size, and statistical analyses performed may account for some of the observed inconsistencies.

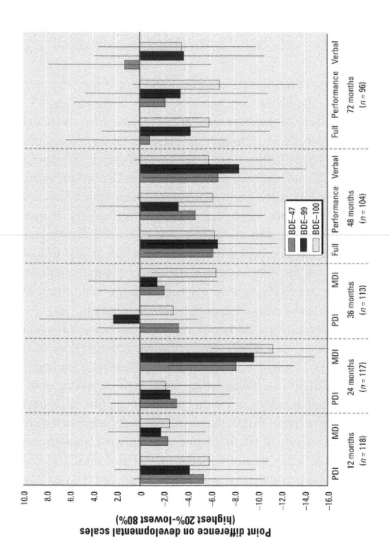

FIGURE 1: Difference in mean developmental score (and 95% confidence interval around the mean) comparing individuals in the highest quintile (20%) of exposure with those in the lower 80% of BDEs 47, 99, and 100. Mean differences were adjusted for age at testing, race/ethnicity, IQ of mother, sex of child, gestational age at birth, maternal age, ETS (yes/no), maternal education, maternal hardship, breast-feeding, language, and location of interview.

Our results are consistent with published toxicologic experiments [reviewed by Costa and Giordano (2007)]. For example, studies evaluating the neurodevelopmental effects of neonatal exposure to PBDEs in mice during critical developmental periods have reported altered habituation patterns (Viberg et al. 2003a, 2003b), hyperactivity (Gee and Moser 2008), and learning and memory deficits (Dufault et al. 2005; Viberg et al. 2003a). There is some evidence suggesting that BDE-99 is more potent than BDE-47 (Viberg et al. 2003a) and also that effects worsen (or are more apparent) with age (Viberg et al. 2003a). In general, we observed the largest associations with prenatal exposure to BDE-100, and the associations with prenatal exposures were still apparent, albeit not consistently significant, in our smaller sample examined at 72 months of age.

A number of potential mechanisms have been proposed to explain the cognitive and locomotive deficits observed in animals after PBDE exposure during critical developmental periods, including direct neurotoxic effects on neuronal and glial cells (Costa et al. 2008) resulting from changes in the quantity of cholinergic nicotinic receptors in the hippocampus (Viberg et al. 2003a) and induction of apoptotic cerebellar granule cell death (Reistad et al. 2006). In addition, there is compelling experimental and epidemiologic evidence suggesting that PBDEs can interfere with thyroid hormone pathways (Legler 2008). Because thyroid hormones are critical for normal brain development, this provides an attractive explanation for observed neurodevelopmental effects after neonatal PBDE exposure (Bigsby et al. 1999; Porterfield 2000). Toxicologic evidence corroborating this theory includes potentially causal associations between neonatal exposure to BDE-47, BDE-99, or commercial PBDE mixtures (DE-71 and Bromokal 70-5 DE) and reduced thyroxine (T_4) concentrations in experimental murine models (Fowles et al. 1994; Hallgren et al. 2001; Kuriyama et al. 2007).

Although only limited human epidemiologic data are available, increased levels of BDEs 47, 99, and 100 in dust in the homes of adult human males recruited through a U.S. infertility clinic were associated with altered hormone levels. PBDEs were inversely associated with free androgen index and with luteinizing and follicle-stimulating hormones and were positively associated with inhibin B, sex hormone–binding globulin, and free T4 (Meeker et al. 2009). In another study of adult males, increased serum concentrations of PBDEs were positively related to T_4 and inversely related to total triiodothyronine (T_3) and thyroid-stimulating hormone (TSH) (Turyk et al. 2008). The positive associations between PBDEs and T_4 levels demonstrated in these human studies are not consistent with the results from experimental animal models, raising the possibility that the underlying mechanism of the effect of PBDEs on thyroid disruption may differ among species. However, it is difficult to extrapolate findings from studies evaluating exposure effects in adults to prenatal exposures, because PBDEs may exhibit differential effects on thyroid hormone levels at different stages of the life span. A recent study of PBDEs measured in human cord blood of infants born to a cross-section of women delivering in Baltimore, Maryland, showed a consistent nonsignificant negative association with both total and free T_4 in infants (Herbstman et al. 2008). More research is necessary to fully characterize the association of human prenatal exposure to PBDEs with thyroid hormone levels.

The exact mechanism of thyroid disruption by PBDEs in humans has not yet been elucidated, but two potential pathways through which PBDE exposure could lead to thyroid disruption have been proposed [reviewed by Zhang et al. (2008)]. The structural similarities of T_4 and T_3 to polyhalogenated aromatic hydrocarbons suggest that hydroxylated PBDE metabolites could displace thyroid hormones from thyroid transport proteins (i.e., transthyretin), altering free thyroid hormone levels (Turyk et al. 2008). Alternatively (or in addition), PBDEs might affect hormone levels by influencing thyroid hormone synthesis and/or stimulating thyroid hormone metabolism (Szabo et al. 2009; Turyk et al. 2008). Brain development in the fetus is contingent on the precise timing of thyroid hormone levels, particularly for T_4, and deviations above or below the normal levels can lead to developmental deficits (Williams 2008). The fetus originally derives all thyroid hormone from the mother, but over the course of the pregnancy, its thyroid gland develops, and hormones produced within the fetus gradually replace the maternal source. The surge in maternal T_4 in the first trimester, coupled with TSH inhibition, is thought to provide a supply of hormone during this critical developmental period, and alteration of T_4 levels by PBDEs at this time could alter neurodevelopment (Williams 2008). Although low serum T4 from maternal hypothyroidism during gestation (e.g., iodine deficiency) is known to cause mental retardation in children, elevated levels of T_4 have been associated with increased rates of miscarriage (Anselmo et al. 2004) and could potentially be linked to neurodevelopmental problems.

Because of their similar chemical structures, PBDEs and polychlorinated biphenyls (PCBs) have been compared in terms of their potential health effects. Although PCBs were banned in most industrialized countries > 25 years ago, they are still measurable in human and environmental samples because of their long half-lives in the environment and in humans (Talsness 2008). Prenatal exposure to PCBs has been shown in several cohort studies to significantly reduce cognitive function during childhood [reviewed by Schantz et al. (2003)] and has also been associated in some studies with altered thyroid hormone levels (Chevrier et al. 2007; Herbstman et al. 2008). Because of the structural similarity of PCBs and PBDEs, it has been postulated that they exert biological effects through similar processes.

This study population is unique in that participants were initially recruited to measure the extent and the effects of prenatal exposure to contaminants (including PBDEs) that were potentially released by the destruction of the WTC towers. Studies examining environmental samples collected pre- and post-9/11 near the WTC site found indications of higher concentrations of PBDEs after the attacks (Litten et al. 2003) and nearer to the WTC disaster site (Butt et al. 2004). These trends may be attributable to debris containing office equipment known to be treated with PBDEs (de Wit 2002). In our study population, cord plasma levels of PBDEs were not significantly related to residential distance from the WTC site. There is some evidence suggesting that PBDE exposure may be related to the WTC attack based on the gestational age on 9/11, such that women who were in the second half of their pregnancy on 9/11 had children with higher cord concentrations of PBDEs (unpublished data). It is also possible that just after 9/11, some women had elevated levels of PBDEs but that these levels declined with the passage of time between the peak exposure and delivery, resulting in lower observed levels. In either scenario, it is not clear how much this apparent association between gestational age on 9/11 and exposure concentration contributes to the body burden, and it is certain that sources other than the WTC are also accountable. In this report, our interest is in the association between the integrated prenatal PBDE exposure from multiple sources and neurodevelopment. It is also possible that there are other unknown factors associated with PBDEs that may confound the observed relationships between prenatal PBDE exposure and adverse neurodevelopment.

Levels of cord blood PBDEs in our population are consistent with those reported in other U.S. populations (Herbstman et al. 2007; Mazdai et al. 2003; Wu et al. 2007). Compared with cord blood measurements in an inner-city population in Baltimore, Maryland, our study population had slightly lower median concentrations (i.e., 11.2 ng/g lipid vs. 13.6 ng/g lipid for BDE-47) (Herbstman et al. 2007). In the Baltimore cohort as well as in this New York City cohort, higher cord PBDE concentrations were associated with mothers' African American or non-Asian race/ethnicity, although a higher proportion of the Baltimore population was African American (70% vs. 15%) and a lower proportion was Asian (8% vs. 30%). Increasing maternal age was associated with lower PBDE concentrations in the Baltimore cohort but not in New York City; however the median maternal age was also lower in Baltimore (25 years vs. 30 years) (Herbstman et al. 2007). The demographic differences between these two populations may explain the small differences in blood levels observed at the population level.

Although dietary ingestion was once thought to be the largest route of PBDE exposure in humans, the similarity of PBDE levels in foods in Europe, Asia, and North America fails to adequately explain the high blood levels in the U.S. population (Frederiksen et al. 2009). Dust inhalation may be a more important exposure route to PBDEs, particularly BDEs 47, 99, and 100. In a review of median PBDE levels in dust and air samples, measured BDE-47 dust levels in Europe and North America were 32 and 429 ng/g of dust, respectively. Similar disparities were observed for BDE-99 and BDE-100 levels (Frederiksen et al. 2009; Sjodin et al. 2008a). Particular attention should be given to this exposure route in young children, who are more likely to encounter dust because of their proximity to the floor. Dust is estimated to contribute from 80 to 93% of PBDE exposure in toddlers, and their small body size compounds the effect of their exposures (Costa and Giordano 2007). In this study, we were not able to control for postnatal dust exposure.

In the general population, infants and toddlers have the highest body burden of PBDEs, and along with dust exposure, exposure via breast milk is thought to be a major contributor to this burden (Costa et al. 2008; Toms et al. 2009). Breast-fed infants are estimated to be exposed to 306 ng/kg body weight/day PDBE compared with 1 ng/kg body weight/day in adults, with the most prominent congeners being BDEs 47, 99, and 153 (Costa et al. 2008). In our study, breast-feeding rates were higher in children with higher cord PBDE levels, indicating that PBDEs measured in cord blood may underestimate the exposure of breast-fed children. Breast-feeding was, as expected, associated with higher scores on neurodevelopmental indices, making it an important potential confounder to include in multivariate statistical models.

12.4 CONCLUSIONS

This report is among the first epidemiologic studies to demonstrate inverse associations between elevated cord blood concentrations of PBDEs and adverse neurodevelopmental test scores. These findings indicate a need for additional work to advance our understanding of the effects of perinatal exposure to PBDEs on neurodevelopment and to evaluate the role of thyroid hormones in this process. Additional PBDE congeners not measured in our study should also be examined to determine whether other congeners, including those that are highly brominated, play a role in developmental outcomes. Future work should also explore the possibility of interactions of PBDEs with other chemicals such as PCBs and dichlorodiphenyldichloroethylene. Although additional studies exploring the associations between PBDE exposure and developmental effects are underway, the identification of opportunities to reduce exposure to these compounds should be a priority.

REFERENCES

1.   Anselmo J, Cao D, Karrison T, Weiss RE, Refetoff S. 2004. Fetal loss associated with excess thyroid hormone exposure. JAMA 292(6):691–695.

2. Bernert JT Jr, Turner WE, Pirkle JL, Sosnoff CS, Akins JR, Waldrep MK, et al. 1997. Development and validation of sensitive method for determination of serum cotinine in smokers and nonsmokers by liquid chromatography/atmospheric pressure ionization tandem mass spectrometry. Clin Chem 43(12):2281–2291.

3. Bigsby R, Chapin RE, Daston GP, Davis BJ, Gorski J, Gray LE, et al. 1999. Evaluating the effects of endocrine disruptors on endocrine function during development. Environ Health Perspect 107: suppl 4):613–618.

4. Brown L, Sherbenou R, Johnson S. 1990. Test of Non-Verbal Intelligence: A Language-Free Measure of Cognitive Ability. Austin, TX:PRO-ED, Inc.

5. Butt CM, Diamond ML, Truong J, Ikonomou MG, Helm PA, Stern GA. 2004. Semivolatile organic compounds in window films from lower Manhattan after the September 11th World Trade Center attacks. Environ Sci Technol 38(13):3514–3524.

6. Chevrier J, Eskenazi B, Bradman A, Fenster L, Barr DB. 2007. Associations between prenatal exposure to polychlorinated biphenyls and neonatal thyroid-stimulating hormone levels in a Mexican-American population, Salinas Valley, California. Environ Health Perspect 115:1490–1496.

7. Costa LG, Giordano G. 2007. Developmental neurotoxicity of polybrominated diphenyl ether (PBDE) flame retardants. Neurotoxicology 28(6):1047–1067.

8. Costa LG, Giordano G, Tagliaferri S, Caglieri A, Mutti A. 2008. Polybrominated diphenyl ether (PBDE) flame retardants: environmental contamination, human body burden and potential adverse health effects. Acta Biomed 79(3):172–183.

9. Darnerud PO. 2008. Brominated flame retardants as possible endocrine disruptors. Int J Androl 31(2):152–160.

10. Darnerud PO, Eriksen GS, Johannesson T, Larsen PB, Viluksela M. 2001. Polybrominated diphenyl ethers: occurrence, dietary exposure, and toxicology. Environ Health Perspect 109: suppl 1):49–68.

11. de Wit CA. 2002. An overview of brominated flame retardants in the environment. Chemosphere 46(5):583–624.

12. Dufault C, Poles G, Driscoll LL. 2005. Brief postnatal PBDE exposure alters learning and the cholinergic modulation of attention in rats. Toxicol Sci 88(1):172–180.

13. Fowles JR, Fairbrother A, Baecher-Steppan L, Kerkvliet NI. 1994. Immunologic and endocrine effects of the flame-retardant pentabromodiphenyl ether (DE-71) in C57BL/6J mice. Toxicology 86(1–2):49–61.

14. Frederiksen M, Vorkamp K, Thomsen M, Knudsen LE. 2009. Human internal and external exposure to PBDEs—a review of levels and sources. Int J Hyg Environ Health 212(2):109–134.

15. Gee JR, Moser VC. 2008. Acute postnatal exposure to brominated diphenylether 47 delays neuromotor ontogeny and alters motor activity in mice. Neurotoxicol Teratol 30(2):79–87.

16. Hallgren S, Sinjari T, Hakansson H, Darnerud PO. 2001. Effects of polybrominated diphenyl ethers (PBDEs) and polychlorinated biphenyls (PCBs) on thyroid hormone and vitamin A levels in rats and mice. Arch Toxicol 75(4):200–208.

17. Herbstman JB, Sjodin A, Apelberg BJ, Witter FR, Halden RU, Patterson DG, et al. 2008. Birth delivery mode modifies the associations between prenatal polychlorinated biphenyl (PCB) and polybrominated diphenyl ether (PBDE) and neonatal thyroid hormone levels. Environ Health Perspect 116:1376–1382.

18. Herbstman JB, Sjödin A, Kurzon M, Needham LL, Patterson DGJ, Wang R, et al. 2007. PBDE exposure, placental transfer and birth outcomes. Epidemiology 18(5):S159..

19. Hites RA. 2004. Polybrominated diphenyl ethers in the environment and in people: a meta-analysis of concentrations. Environ Sci Technol 38(4):945–956.

20. Hovander L, Athanasiadou M, Asplund L, Jensen S, Wehler EK. 2000. Extraction and cleanup methods for analysis of phenolic and neutral organohalogens in plasma. J Anal Toxicol 24(8):696–703.

21. Jedrychowski W, Perera F, Mroz E, Edwards S, Flak E, Bernert JT, et al. 2009. Fetal exposure to secondhand tobacco smoke assessed by maternal self-reports and cord blood cotinine: prospective cohort study in Krakow. Matern Child Health J 13(3):415–423.

22. Kaplan C.. 1993. Predicting first-grade achievement from pre-kindergarten WPPSI-R scores. J Psychoeduc Assess 11(2):133–138.

23. Kuriyama SN, Wanner A, Fidalgo-Neto AA, Talsness CE, Koerner W, Chahoud I. 2007. Developmental exposure to low-dose PBDE-99: tissue distribution and thyroid hormone levels. Toxicology 242(1–3):80–90.

24. Lederman SA, Jones RL, Caldwell KL, Rauh V, Sheets SE, Tang D, et al. 2008. Relation between cord blood mercury levels and early child development in a World Trade Center cohort. Environ Health Perspect 116:1085 1091.

25. Lederman SA, Rauh V, Weiss L, Stein JL, Hoepner LA, Becker M, et al. 2004. The effects of the World Trade Center event on birth outcomes among term deliveries at three lower Manhattan hospitals. Environ Health Perspect 112:1772–1778.

26. Legler J.. 2008. New insights into the endocrine disrupting effects of brominated flame retardants. Chemosphere 73(2):216 222.

27. Litten S, McChesney DJ, Hamilton MC, Fowler B. 2003. Destruction of the World Trade Center and PCBs, PBDEs, PCDD/Fs, PBDD/Fs, and chlorinated biphenylenes in water, sediment, and sewage sludge. Environ Sci Technol 37(24):5502–5510.

28. Mazdai A, Dodder NG, Abernathy MP, Hites RA, Bigsby RM. 2003. Polybrominated diphenyl ethers in maternal and fetal blood samples. Environ Health Perspect 111:1249–1252.

29. Meeker JD, Johnson PI, Camann D, Hauser R. 2009. Polybrominated diphenyl ether (PBDE) concentrations in house dust are related to hormone levels in men. Sci Total Environ 407(10):3425–3429.

30. Porterfield SP. 2000. Thyroidal dysfunction and environmental chemicals—potential impact on brain development. Environ Health Perspect 108: suppl 3):433–438.

31. Qiu X, Bigsby RM, Hites RA. 2009. Hydroxylated metabolites of polybrominated diphenyl ethers in human blood samples from the United States. Environ Health Perspect 117:93–98.

32. Reistad T, Fonnum F, Mariussen E.. 2006. Neurotoxicity of the pentabrominated diphenyl ether mixture, DE-71, and hexabromocyclododecane (HBCD) in rat cerebellar granule cells in vitro. Arch Toxicol 80(11):785–796.

33. Roze E, Meijer L, Bakker A, Van Braeckel KNJA, Sauer PJJ, Bos AF. 2009. Prenatal exposure to organohalogens, including brominated flame retardants, influences motor, cognitive, and behavioral performance at school age. Environ Health Perspect 117:1953–1958.

34. Schantz SL, Widholm JJ, Rice DC. 2003. Effects of PCB exposure on neuropsychological function in children. Environ Health Perspect 111:357–376.

35. Sjodin A, Jones RS, Lapeza CR, Focant JF, McGahee EE 3rd, Patterson DG Jr. 2004. Semiautomated high-throughput extraction and cleanup method for the measurement of polybrominated diphenyl ethers, polybrominated biphenyls, and polychlorinated biphenyls in human serum. Anal Chem 76(7):1921–1927.

36. Sjodin A, Papke O, McGahee E, Focant JF, Jones RS, Pless-Mulloli T, et al. 2008. . Concentration of polybrominated diphenyl ethers (PBDEs) in household dust from various countries. Chemosphere 73: suppl 1):S131–S136.

37. Sjodin A, Wong LY, Jones RS, Park A, Zhang Y, Hodge C, et al. 2008. . Serum concentrations of polybrominated diphenyl ethers (PBDEs) and polybrominated biphenyl (PBB) in the United States population: 2003–2004. Environ Sci Technol 42(4):1377–1384.

38. Szabo DT, Richardson VM, Ross DG, Diliberto JJ, Kodavanti PR, Birnbaum LS. 2009. Effects of perinatal PBDE exposure on hepatic phase I, phase II, phase III, and deiodinase 1 gene expression involved in thyroid hormone metabolism in male rat pups. Toxicol Sci 107(1):27–39.

39. Talsness CE. 2008. Overview of toxicological aspects of polybrominated diphenyl ethers: a flame-retardant additive in several consumer products. Environ Res 108(2):158–167.

40. Toms LL, Sjodin A, Harden F, Hobson P, Jones R, Edenfield E, et al. 2009. Serum polybrominated diphenyl ether (PBDE) levels are higher in children (ages 2 to 5 years) than in infants and adults. Environ Health Perspect 117:1461–1465.

41. Tong IS, Lu Y. 2001. Identification of confounders in the assessment of the relationship between lead exposure and child development. Ann Epidemiol 11(1):38–45.

42. Turyk ME, Persky VW, Imm P, Knobeloch L, Chatterton R, Anderson HA. 2008. Hormone disruption by PBDEs in adult male sport fish consumers. Environ Health Perspect 116:1635–1641.

43. Viberg H, Fredriksson A, Eriksson P.. 2003. . Neonatal exposure to polybrominated diphenyl ether (PBDE 153) disrupts spontaneous behaviour, impairs learning and memory, and decreases hippocampal cholinergic receptors in adult mice. Toxicol Appl Pharmacol 192(2):95–106.

44. Viberg H, Fredriksson A, Jakobsson E, Orn U, Eriksson P.. 2003. . Neurobehavioral derangements in adult mice receiving decabrominated diphenyl ether (PBDE 209) during a defined period of neonatal brain development. Toxicol Sci 76(1):112–120.

45. Williams GR. 2008. Neurodevelopmental and neurophysiological actions of thyroid hormone. J Neuroendocrinol 20(6):784–794.

46. Wu N, Herrmann T, Paepke O, Tickner J, Hale R, Harvey LE, et al. 2007. Human exposure to PBDEs: associations of PBDE body burdens with food consumption and house dust concentrations. Environ Sci Technol 41(5):1584–1589.

47. Zhang Y, Guo GL, Han X, Zhu C, Kilfoy BA, Zhu Y, et al. 2008. Do polybrominated diphenyl ethers (PBDEs) increase the risk of thyroid cancer? Biosci Hypotheses 1(4):195–199.

There are three tables as well as several supplemental files that are not available in this version of the article. To view this additional information, please use the citation on the first page of this chapter.

CHAPTER 13

NEUROBEHAVIORAL FUNCTION AND LOW-LEVEL EXPOSURE TO BROMINATED FLAME RETARDANTS IN ADOLESCENTS: A CROSS-SECTIONAL STUDY

MICHAŁ KICINSKI, MINEKE K. VIAENE, ELLY DEN HOND, GREET SCHOETERS, ADRIAN COVACI, ALIN C. DIRTU, VERA NELEN, LIESBETH BRUCKERS, KIM CROES, ISABELLE SIOEN, WILLY BAEYENS, NICOLAS VAN LAREBEKE, AND TIM S. NAWROT

13.1 BACKGROUND

Brominated flame retardants (BFR) are chemicals widely used in a variety of household and commercial products including plastics, electric equipment, textiles, and polyesters in order to prevent fire [1,2]. Many of them bioaccumulate in the environment and have been found in water, air, biota, human tissues, breast milk, and blood [3-6]. House dust and food represent two important sources of human exposure [5,7].

Neurobehavioral Function and Low-Level Exposure to Brominated Flame Retardants in Adolescents: A Cross-Sectional Study. Kicinski M, Viaene MK, Den Hond E, Schoeters G, Covaci A, Dirtu AC, Nelen V, Bruckers L, Croes K, Sioen I, Baeyens W, Van Larebeke N, and Nawrot TS. Environmental Health, *11,86 (2012), doi:10.1186/1476-069X-11-86. Licensed under Creative Commons Attribution 2.0 Generic License, http://creativecommons.org/licenses/by/2.0/.*

A number of animal studies showed effects of a prenatal and postnatal exposure to BFR on neurodevelopment and were recently reviewed. [8-10] Neurobehavioral effects during juvenile development or adulthood have been observed in rodents after a brief postnatal exposure to polybrominated diphenyl ethers (PBDE) 47 [11,12], 99 [11,13-16], 153 [17], 203 [18], 206 [18], 209 [19-22], the commercial PBDE mixture DE-71 [23,24], and hexabromocyclododecane (HBCD) [25,26], a chronic perinatal exposure to PBDE-47 [27,28] and PBDE-99 [29], and an acute prenatal exposure to PBDE-99 [30]. Detrimental effects of PBDE exposure on neurodevelopment have also been reported in zebrafish [31,32]. Changes in the motor activity have been most frequently studied and best documented [8,10]. Also in vitro studies support the hypothesis of neurotoxicity of BFR. PBDE congeners were capable of inducing oxidative stress [33-35] and apoptosis [33,34] in cultured neurons.

Despite the fact that the first results of the experimental animal studies suggesting a neurotoxic potential of BFR were available more than 10 years ago, the effects in humans have not been extensively investigated to date. Three small prospective studies [36-38] evaluated the effects of a perinatal exposure to BFR on neurobehavioral function in children. Concentrations of several PBDE congeners in umbilical cord blood of newborns showed an association with indicators of neurodevelopment in early childhood. [37] In another study [36], consistent neurodevelopmental effects at the age of 8–12 months of the exposure measured by breast milk PBDE concentrations were not observed. Roze et al. [38] reported both negative and positive neurobehavioral effects of a prenatal exposure to HBCD and several PBDE congeners among 5–6 year-old children. In a cross-sectional study on older adults [39], PBDE's measured in the serum were not associated with performance in cognitive tasks. All associations reported in these studies were investigated using less than 150 participants.

The thyroid system is one of the targets of BFR [40]. Experimental animal studies have demonstrated that a PBDE exposure may result in a decrease of blood thyroxine [22,24,41-45] and triiodothyronine [43,44,46-48] levels. These effects were observed not only in gestation and early childhood, but also later in life [41,42]. A disruption of the thyroid system has been suggested as a mediator of the BFR neurotoxicity [9,49].

We conducted a cross-sectional study of the association between neurobehavioral function and biomarkers of exposure to BFR [serum levels of polybrominated diphenyl ether (PBDE) congeners 47, 99, 100, 153, 209, hexabromocyclododecane (HBCD), and tetrabromobisphenol A (TBB-PA)] in a group of Flemish adolescents. Additionally, we investigated the association between BFR and the thyroid function as a potential biological mechanism responsible for the neurotoxicity of these chemicals.

13.2 METHODS

13.2.1 STUDY POPULATION AND DATA COLLECTION

The study was a part of a biomonitoring program for environmental health surveillance in Flanders, Belgium. The participants were recruited between 2008 and 2011 in two industrial areas (Genk and Menen) and from the general population of Flemish adolescents. Participants were eligible if they studied in the third year of secondary school. Hence, most participants were 14 or 15 year old, but older students were also allowed in the study.

In the general Flemish population, random sampling was attained through a multistage sampling design. In the first step, ten schools (two in each of the five Flemish provinces)—at least 20 km apart from each other—were randomly selected. In the second step, classes were randomly selected within each school and all pupils in a class were invited until the provided number of participants was reached. The number of participants per province was proportional to the number of inhabitants in that province (status at 01/01/2006).

In Genk and Menen, study areas were defined based on environmental data and the location of the industrial sites. All pupils in the requested age range living within the selected study area were eligible. Names and addresses were attained from the population registry. In Genk 54% of the adolescents were invited via a letter send to the home address and 46% during a home visit. In Menen, all participants were invited via a letter send to the home-address. Due to a small response, 30% of the invited children were contacted again via schools and 9% via a home visit.

TABLE 1: Descriptive statistics

N=515[A]	
Boys	271 (52.6%)
Age, years	14.9 (0.7)
BMI, kg/m^2	20.4 (3.1)
Type of education – general secondary	290 (56.3%)
The highest level of education of parents	
no diploma	22 (4.3%)
9 grades	52 (10.1%)
12 grades	163 (31.7%)
College of university diploma	278 (54%)
Parents owning the house	456 (88.5%)
Current smoking	65 (12.6%)
Passive smoking[B]	82 (15.9%)
Alcohol use at least once a month	129 (25%)
Number of hours a week using computer	
< 2	50 (9.7%)
2-9	271 (52.6%)
10-19	147 (28.5%)
≥ 20	47 (9.1%)
Fish consumption[C]	
Low	224 (43.5%)
Average	186 (36.1%)
High	105 (20.4%)
Physical activity in leisure time at least once a week	405 (78.6%)
Blood lipids, mg/dl	448.9 (72.9)
Blood lead, µg/dl	1.4 (0.7 to 2.9)
Sum of serum PCB 138, 153 and 180, ng/L	171.6 (58 to 445)
Thyroid hormones serum levels	
FT3, pg/mL	4.15 (0.53)
FT4, ng/dL	1.24 (0.17)
TSH, µU/mL	2.15 (0.99 to 4.45)
Neurobehavioral parameters	
Continuous Performance, reaction time, msec, N=489	409.2 (41.8)
Continuous Performance, errors of omission, N=489	2.3 (2.7)
Continuous Performance, errors of commission, N=489	5.6 (3.5)

TABLE 1: *Cont.*

N=515[A]	
Digit Symbol, sec, N=340	98.3 (17.7)
Digit Span Forward, N=511	5.6 (1.03)
Digit Span Backward, N=499	4.49 (1.01)
Finger Tapping, preferred hand, N=511	293.7 (40.2)
Finger Tapping, non-preferred hand, N=509	258.6 (33.8)

Arithmetic mean (standard deviation) is given for the continuous variables used on their original scale and geometric mean (5th to 95th percentile) for the logarithmically transformed continuous variables. Count (percent) is given for the categorical variables.
[A]*Participants for whom information about the covariates, brominated flame retardants blood levels, and at least 1 neurobehavioral measure was available.*
[B]*At least 1 family member smoking inside the house.*
[C]*Based on two questions about the amount of warm and cold fish eaten per week. Low corresponds to less than 20 g, middle to 20–25 g and high to more than 25 g per day.*

Two weeks before the study session, subjects received two questionnaires to fill in, one for themselves and one for their parents. The questionnaire for adolescents included information about their exercising habits, amount of time spent using a computer, alcohol use, and smoking. Questions about the socioeconomic status, passive smoking and eating habits were included in the questionnaire for parents. The study session including an administration of the neurological tests, a collection of a blood sample, and a measurement of the length and the weight was around 1 hour long. Each subject received a 10 Euro voucher for the participation. Both parents and teenagers provided informed consent for participation. The study was approved by the Ethical Committee of the University of Antwerp.

The response rate equaled 22.1% in the general Flemish population, 34.3% in Genk and 22.5% in Menen. A non-responder analysis performed in a group of 106 adolescents (30 participants and 76 non-participants) did not reveal differences in socio-economic status indicators type of education (general secondary education versus other, $p=0.58$), education of the father (higher education vs not, $p=0.99$), or education of the mother (higher education vs not, $p=0.22$) between participants and non-participants. The proportion of girls was higher among the participants (83.3%

vs. 61.8% in non-participants, p=0.03), but an equal distribution between boys and girls was a stratification criterion in the recruitment strategy. 606 adolescents participated in the study. Blood measurements were not available for three participants, and four participants did not complete any of the neurobehavioral tests. Additionally, for 84 out of the remaining 599 participants information on at least one of the covariates used in the analysis was missing. 515 subjects who completed at least one neurobehavioral test and for whom information about the covariates and serum BFR levels was available, were used in the analysis (see Table 1). This group consisted of 163 adolescents from Genk, 178 from Menen, and 174 from the general Flemish population.

13.2.2 NEUROBEHAVIORAL TESTS

Neurobehavioral Evaluation System (NES) is a computerized battery of tests developed to study the neurological effects of an exposure to environmental agents [50]. NES has been used in a number of studies investigating the neurobehavioral impact of neurotoxicants and dose–response relationships with intensity of exposure were reported [51]. In our study, we used four tests from the NES-3 version of the test battery [52] (see Figure 1).

In the Continuous Performance test, a series of letters is displayed on the screen, one at a time, and each for approximately 200 msec. The task is to immediately respond to the letter S, and not to other letters, by pressing the spacebar. A new letter appears each 1000 msec. In total, the letter S appears 60 times. The mean reaction time for responding to the target letter in msec, the number of errors of omission, i.e., the number of times that a subject did not react within 1200 msec, the number of errors of commission, i.e., the number of responses when the letter S was not displayed, were used as the measure of performance. The test evaluates sustained attention. It showed a good test-retest reliability in a group of patients directed to a neuropsychological examination [51].

In the Digit-Symbol test, a row of 9 symbols paired with 9 digits is displayed at the top of the screen. The same 9 symbols but in a different order are shown at the bottom. When a digit is displayed, the task is to indicate the symbol, which is paired with this digit, from the bottom row. A new digit appears only after the correct symbol has been indicated. In total, 27 digits are displayed. The total time needed to complete the test measured in seconds describes the performance. The test is characterized by a satisfactory reliability [52]. In a part of the study area, a different test was administered instead of the Digit-Symbol test. As a result, the results of this test were only available for a group of 341 participants from Genk and Menen.

The Digit Span test consists of two parts. In the first part, a subject hears a sequence of digits. The task is to reproduce them. In case of a correct answer, a one digit longer sequence is presented. In case of a mistake, a sequence of the same length is presented. When two incorrect answers

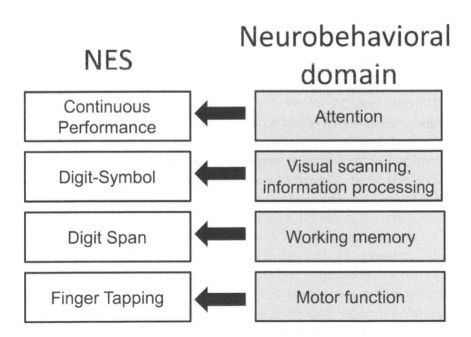

FIGURE 1: An overview of the neurological tests used in the study.

in a row are given, the first part of the test finishes. The second part is the same as the first one, but the sequences are reproduced in the reverse order. Digit Span Forward is the maximum span reproduced in the first part. Digit Span Backward is the maximum span reproduced in the reverse order. The first part of the test assesses the working memory span. Good performance in the second part requires both the ability to hold and manipulate information. In a part of the study area, the Digit Span test was administered using computers with touch screens. In this case the task was to indicate the digits on the screen, not using the keyboard. In order to account for a possible effect of the way the test was administered, an indicator variable was included in the regression models for this test.

In the Finger Tapping test a subject presses the spacebar as many times as possible during a trial of 10 sec. The first part of the test consists of 4 trials with the preferred-hand. The second part consists of 4 trials with the non-preferred hand. The summary measures are the total number taps with the preferred-hand and the total number of taps with the non-preferred hand. The test measures the manual motor speed.

13.2.3 BLOOD SAMPLES ANALYSIS

PBDE congeners 28, 47, 99, 100, 153, 154, 183, and 209, HBCD, and tetrabromobisphenol A (TBBPA) were measured in the serum according to the method described by Covaci and Voorspoels [53]. Briefly, solid phase extraction was performed to prepare the samples. The eluate was purified on acid silica. The extract was further analyzed by gas chromatography mass spectrometry in electron capture negative ion mode using a 25 m × 0.22 mm × 0.25 μm HT-8 column.

Characteristics of the distributions of the BFR are shown in Table 2. PBDE congeners 28, 154 and 183 for which less than 5% of the observations had a concentration higher than the limit of quantification (LOQ), were not used in the analysis. For PBDE 47, 99, 100, 153, and 209 binary exposure indicators were used with the LOQ values as thresholds. The logarithm of the sum of PBDE 47, 99, 100, 153 was used as a measure of the total long-term PBDE exposure. In the calculation of the sum, the values below LOQ were replaced by LOQ/2.

TABLE 2: Concentrations of polybrominated flame retardants in serum (ng/L)

	LOQ	Median	P75	P95	Max
BDE28	2	<LOQ	<LOQ	<LOQ	24
BDE47	3	<LOQ	3	9	104
BDE99	3	<LOQ	<LOQ	3	12
BDE100	2	<LOQ	<LOQ	2	42
BDE153	2	2	3	8	24
BDE154	2	<LOQ	<LOQ	<LOQ	6
BDE183	2	<LOQ	<LOQ	<LOQ	5
BDE209	25	<LOQ	<LOQ	53	325
HBCD	30	<LOQ	<LOQ	59	234
TBBPA	15	<LOQ	<LOQ	22	186
SUM PBDE[A]		7	10	21	125

LOQ – limit of quantification.
[A]*The sum of PBDE congeners 47, 99, 100, and 153. In the calculation, values lower than LOQ were replaced by LOQ/2.*

Concentrations of PCB congeners in the serum were determined using the same method as for the BFR. The sum of PCB 138, 153, and 180 transformed logarithmically was used as an indicator of the PCB exposure. Lead was measured in the whole blood as described by Schroijen et al. [54]. Blood lipids were measured graphimetrically. Thyroid hormones FT3, FT4 and TSH were measured by competitive immune assays.

13.2.4 STATISTICAL ANALYSIS

We used SAS software version 9.2 (SAS Institute Inc, Cary, NC) for all analysis. Continuous positive-value variables with a right-skewed distribution were logarithmically transformed. Normal quantile plots of the residuals were used to examine the normality assumption for all linear models. For the individual PBDE congeners, HBCD and TBBPA the effects of the concentrations above the LOQ compared to the concentrations below the LOQ were estimated. The sum of PBDE's was transformed logarithmically, and the effect of its two-fold increase was estimated.

In all the models investigating the effects on the neurobehavioral parameters, we corrected for age of a child, gender, type of education (general secondary education versus other), the highest level of education attained by either of the parents (using three indicator variables), whether or not the parents owned the house, smoking, passive smoking, and blood lipids. Models evaluating the effects on the number of digits reproduced in the Digit Span test were also corrected for the method of test administration (touch screen versus keyboard). Additionally, we corrected for the covariates BMI of a child, physical activity in leisure time at least once a week, computer use, alcohol use at least once a month, fish consumption, the logarithm of blood lead and the logarithm of serum PCB's 138, 153 and 180, which were selected using a stepwise regression procedure with $p=0.15$ for entering and $p=0.10$ for remaining in the model. In the models investigating the effects of BFR on FT3, FT4, and TSH levels, we corrected for age, gender, BMI, and blood lipids in all models. Other variables mentioned above besides computer use were included in the model based on a stepwise regression procedure with $p=0.15$ for entering and $p=0.10$ for remaining in the model. Finally, we investigated whether the effects of BFR were modified by gender by including the interaction term in the regression model.

13.3 RESULTS

13.3.1 CHARACTERISTICS OF THE STUDY POPULATION

The characteristics of the study group are given in Table 1. The adolescents (52.6% boys) were between 13.6 and 17 years of age, and the mean age equaled 14.9 years. A majority of the participants (54%) had at least one parent who graduated from a college or a university. A summary of the results obtained in the neurobehavioral tests is given in Table 1.

TABLE 4: Estimated effects of serum levels of brominated flame retardants on performance in the Continues Performance, Digit-Symbol and Digit Span tests

| | Continuous Performance (N=489) | | | | | | Digit Symbol (N=340) | | Digit Span | | | |
| | Reaction time (msec)^A | | Errors of omission^B | | Errors of commission^B | | Total latency (sec)^A | | Forward^A (N=511) | | Backward^A (N=499) | |
	Effect	95% CI	Effect	95% CI	Effect	95% CI	Effect	95% CI	Effect	95% CI	Effect	95% CI
PBDE47	3.45	-4.88 to 11.78	-10%	-29.9 to 15.6%	6.2%	-6.2 to 20.1%	-1.19	-5.57 to 3.2	-0.09	-0.29 to 0.11	-0.07	-0.27 to 0.14
PBDE 99	-5.39	-19.85 to 9.08	-16.3%	-46.4 to 30.8%	12.9%	-8.6 to 39.5%	-1.35	-8.94 to 6.24	0.09	-0.25 to 0.44	0.3	-0.04 to 0.64
PBDE100	7.61	-5.94 to 21.16	-5.8%	-37.2 to 41.4%	-3.2%	-21 to 18.5%	1.98	-5.6 to 9.56	-0.26	-0.57 to 0.06	-0.18	-0.49 to 0.14
PBDE153	5.09	-2.76 to 12.95	-19.3%	-36.4 to 2.3%	-2%	-12.8 to 10.2%	-1.34	-5.46 to 2.77	-0.09	-0.28 to 0.1	-0.08	-0.27 to 0.11
PBDE209	-1.2	-14.49 to 12.1	-17.7%	-45.1 to 23.4%	1.4%	-16.9 to 23.7%	2.08	-4.07 to 8.23	0.06	-0.26 to 0.38	-0.26	-0.57 to 0.05
HBCD	-3.53	-18.72 to 11.67	27.8%	-17.5 to 97.9%	21.8%	-2.5 to 52.2%	-0.44	-6.59 to 5.72	0.13	-0.22 to 0.49	-0.04	-0.39 to 0.31
TBBPA	-2.25	-17.28 to 12.77	-9.3%	-43 to 44.2%	-17.7%	-34.7 to 3.9%	-2.48	-10.36 to 5.41	0.03	-0.32 to 0.37	-0.05	-0.41 to 0.3
SUM PBDE	2.12	-2.9 to 7.13	-6.6%	-19.9 to 8.9%	0.7%	-6.6 to 8.6%	-0.39	-3.04 to 2.26	-0.01	-0.13 to 0.11	-0.04	-0.16 to 0.08

TABLE 4: *Cont.*

A A linear model. B A negative binomial model (estimated change in%). For the individual PBDE congeners, HBCD, and TBBPA the effects of levels above the LOQ were estimated. Sum of serum PBDE's 47, 99, 100, and 153 was logarithmically transformed and the effects of its two-fold increase were estimated. All models were adjusted for gender, age, type of education (general secondary education versus other), the highest level of education of parents (using three indicator variables), whether or not the parents owned the house, smoking, passive smoking, and blood lipids. Models evaluating the effects on the number of digits reproduced in the Digit Span test were in addition adjusted for the method of test administration (touch screen verses keyboard). Additionally, BMI, physical activity in leisure time at least once a week, computer use, alcohol use at least once a month, fish consumption, the logarithm of blood lead and the logarithm of serum PCB's 138, 153, and 180 were included in the model based on the stepwise regression procedure with p = 0.15 for entering and p = 0.10 for remaining in the model.

13.3.2 DETERMINANTS OF NEUROBEHAVIORAL FUNCTION

Estimates of the effects of the covariates on the neurobehavioral parameters are shown in Table 3. Gender, age, type of education, parental education, and physical activity were the most important determinants of the performance in the tests.

13.3.3 ASSOCIATIONS BETWEEN BFR AND THE NEUROBEHAVIORAL FUNCTION

We did not find any significant associations between serum levels of BFR and performance in the Continuous Performance, Digit-Symbol or Digit Span tests (Table 4). However, PBDE's were associated with a deterioration of the performance in the Finger Tapping test in the preferred-hand condition (Figure 2). In the continuous analysis, a two-fold increase of the sum of serum PBDE's was associated with a decrease of the number of taps with the preferred-hand by 5.31 (95% CI: 0.56 to 10.05, p=0.029). The model explained 9.85% of the total variability and 0.87% of the variability could be attributed to the sum of serum PBDE's. Concentrations above LOQ were associated with an average decrease of 7.04 taps (95% CI: -0.78 to 14.87; p=0.078) for serum PBDE-47, 12.13 (95% CI:

-1.3 to 25.57; p=0.078) for serum PBDE-99, 12.43 (95% CI: -0.03 to 24.89; p=0.051) for serum PBDE-100, and 8.43 (95% CI: 1.01 to 15.86; p=0.026) for serum PBDE-153. The associations between serum PBDE's and the number of taps with the non-preferred hand were usually consistent (negative association), but did not reach the level of significance. Serum HBCD and TBBPA levels were not significantly associated with performance in the Finger Tapping test.

Also after adjusting for the number of errors of omission and commission, none of the BFR exposure indicators was significantly related with the mean reaction time in the Continuous Performance test. Exposure to BFR did not show negative associations with performance in the Continuous Performance test in analysis stratified by period (Figure 3). Gender did not significantly modify the association between the sum of PBDE's and the number of taps with the preferred hand (p=0.25).

13.3.4 ASSOCIATIONS WITH THE FT3, FT4, AND TSH LEVELS

The estimated associations between BFR and FT3, FT4, and TSH serum levels after correction for possible confounders are shown in Figure 4. Serum levels above LOQ were associated with an average decrease of FT3 level by 0.18 pg/mL (95% CI: 0.03 to 0.34, p=0.020) for PBDE 99 and by 0.15 pg/mL (95% CI: 0.004 to 0.29, p=0.045) for PBDE-100. For the other PBDE congeners the associations had the same direction but were not statistically significant. We did not observe significant associations between PBDE congeners and FT4 levels. PBDE-47 level above LOQ was associated with an average increase of TSH levels by 10.1% (95% CI: 0.8% to 20.2%, p=0.033). The other PBDE congeners were not significantly associated with TSH.

The continuous analyses did not show significant associations between PBDE's and the thyroid hormone levels. For a two-fold increase of the sum of serum PBDE's, FT3 was estimated to decrease by 0.05 pg/mL (95% CI: -0.01 to 010, p=0.10), FT4 to increase by 0.017 ng/dL (95% CI: -0.003 to 0.032, p=0.10), and TSH to increase by 3.9% (95% CI: -1.5% to 9.6%, p=0.16). We did not observe any significant effects of HBCD or TBBPA on the hormone levels. The adjustment for the levels of the thyroid hormones did not substantially change the estimate of the effect of the sum of PBDE's on the number of taps with the preferred-hand.

13.4 DISCUSSION

Consistently with the experimental animal studies demonstrating that exposure to PBDE's during gestation and early childhood affects the motor function [11-22,26-28], we observed negative associations between the serum levels of these BFR and motor speed of the preferred hand in adolescents. The associations between PBDE's and the second indicator of the motor function, the number of taps with the non-preferred hand were not significant but showed the same trend. The non-preferred hand test is performed after the preferred hand test in the Finger Tapping test. Our results resemble observations from the experimental animal studies in which a decrease of the motor activity related to PBDE exposure was present only in the beginning of the tests [11,14-21,25]. We did not observe negative associations between BFR and neurobehavioral domains other than the motor function.

Human data on the neurobehavioral effects of BFR are scarce. In the United States, a prenatal PBDE exposure was inversely associated with the level of mental development at the age of two and inteligence at the age of two and three [37]. In a Dutch study, a prenatal exposure to PBDE-47 and PBDE-99 showed a negative association with sustained attention and PBDE-153 with verbal memory measured at the age of five and six [38]. In contrast to these two studies, we conducted a cross-sectional study and focused on older children. The serum levels of PBDE's were not associated with neurobehavioral outcomes in a cross-sectional study of older adults in New York [39]. To our knowledge, the neurobehavioral effects in adolescents have not been studied yet.

Higher PBDE-99 and PBDE-100 serum levels were significantly associated with a lower level of serum FT3, and the results for the other PBDE congeners showed the same tendency. A negative association between a PBDE exposure and the triiodothyronine concentration was also seen in some other epidemiological studies [55,56]. Consistently with the effects on FT3, most of the indicators of exposure to PBDE's were positively related with TSH levels, although only for PBDE-47 the level of significance was reached.

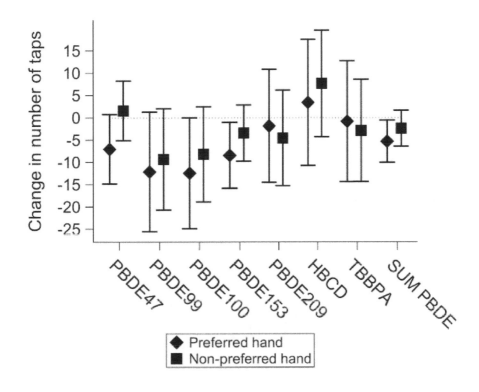

FIGURE 2: Estimated effects of the serum levels of brominated flame retardants on the performance in the Finger Tapping test. For the PBDE congeners, HBCD, and TBBPA the effects of levels above the LOQ were estimated. Sum of PBDE's 47, 99, 100, and 153 was logarithmically transformed and the effects of its two-fold increase were estimated. All models were adjusted for: gender, age, type of education (general secondary education versus other), the highest level of education of parents (using three indicator variables), whether or not the parents owned the house, smoking, passive smoking, and blood lipids. Additionally, BMI, physical activity in leisure time at least once a week, computer use, alcohol use at least once a month, fish consumption, the logarithm of blood lead and the logarithm of serum PCB's 138, 153, and 180 were included in the model based on the stepwise regression procedure with p=0.15 for entering and p=0.10 for remaining in the model.

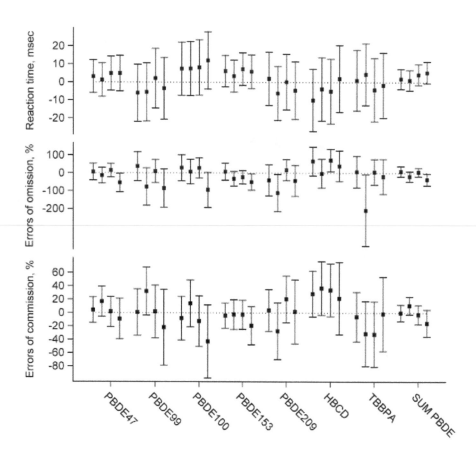

FIGURE 3: Estimated effects of the serum levels of brominated flame retardants on the performance in the Continuous Performance from analysis with stratification by period. For each exposure indicator, the effect on outcome in the first, the second, the third and the fourth block is shown. Each block consisted of 12 trials. The same modeling strategy as in the analysis without stratification was applied.

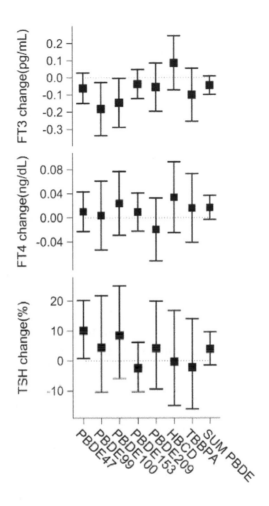

FIGURE 4: Estimated effects of the serum levels of brominated flame retardants on serum FT3, FT4, and TSH concentrations. For the PBDE congeners, HBCD, and TBBPA the effects of levels above the LOQ were estimated. Sum of PBDE's 47, 99, 100, and 153 was logarithmically transformed and the effects of its two-fold increase were estimated. All models were adjusted for: gender, age, BMI and blood lipids. Additionally, type of education (general secondary education versus other), the highest level of education of the parents (using three indicator variables), whether or not the parents own the house, smoking, passive smoking, physical activity, alcohol use, fish consumption, BMI, the logarithm of blood lead and the logarithm of the sum of serum PCB's 138, 153 and 180 were included in the model based on the stepwise regression procedure with p=0.15 for entering and p=0.10 for remaining in the model.

Contrary to the experimental animal studies [22,24,41-45], we did not observe a negative association between PBDE's and FT4 concentration. This is in agreement with other epidemiological studies in humans in which either non-significant associations or significant positive associations were obtained [56-59]. The discrepancy between animal and human data can be possibly explained by high PBDE doses used in the animal studies in which the effects on thyroxine levels were observed. Although we observed a positive association between PBDE-47 and TSH, the FT4 levels were not negatively associated with PBDE's. A possible explanation is that PBDE's may inhibit the deiodinase enzymes which serve to metabolize thyroxine to triiodothyronine [59], resulting in an increase in circulating thyroxine and a decrease in ciruculating triiodothyronine levels.

The biological mechanisms of the effects of PBDE's on the thyroid hormones circulating in the blood have not been fully understood yet. PBDE's exposure caused histological and morphological changes in the thyroid gland in rats indicating its decreased activity [43,60]. It also strongly upregulated uridinediphosphate-glucuronosyltransferase, an enzyme transforming molecules including thyroid hormones into excretable metabolites [44,45]. PBDE exposure also resulted in induction of pentoxy-resorufin-O-deelthylase activity [42,44,45].

Controlling for the thyroid hormone levels did not substantially change the estimated effects of the sum of PBDE's on the motor speed. Besides the effects on the thyroid function, a disturbance of the cholinergic system may be a pathway by which PBDE's affect the motor function. A neonatal PBDE exposure resulted in reduced or hypoactive behavioral responses to cholinergic agonist nicotine [19,61] and a decreased amount of nicotine receptors in adult rodents [17,62]. PBDE's are also capable to disrupt calcium homeostasis in the brain, cause oxidative stress and apoptotic cell death [9,49].

Elimination of PBDE congener from human tissues depends strongly on the level of bromination [63-65]. The half-life time in human tissues has been estimated to be around 2 weeks for PBDE-209 [63-65], between a year and a few years for PBDE-47, PBDE-99, and PBDE-100 [63,65], and may be even longer for PBDE-153 [63,65]. Therefore, serum concentration of PBDE's 47, 99, 100, 153, and the sum of these congeners' concentration indicate to a large extent a long-term exposure. The estimated

HBCD total body half-life time equaled 64 days [63]. Most of TBBPA was excreted from the body of rat within a few days after administration [66]. The serum BFR levels at adolescence which we used as exposure indicators, were unlikely to be strongly affected by the exposure during gestation and the first years of life, which may be a period of a particular susceptibility to the neurotoxicity of BFR.

The main limitation of our study was a large number of observations for which PBDE levels took values below the limit of quantification. In order to deal with this problem, we used binary exposure indicators. However, the limits of quantification which we used to create the categories did not represent critical values separating safe and dangerous exposure levels. This dichotomization of continouos exposure indicators and a possible missclassification due to the use of thresholds which did not have biological relevance might have substantially reduced the statistical power. For 222 participants the levels of both BDE-47, BDE-99, BDE-100, and BDE-153, were lower than the limit of quantification. However, the variability in the sum of these congeners observed in the rest of the participants made it an interesting measure of an overall exposure to PBDE's. Replacing the levels below LOQ with the constant value of LOQ/2 in the calcuation of the sum introduced some measurement error. This error might have lead to an underestimation of the effect of an overall PBDE exposure.

Another disadvantage of our study is that the only aspect of the motor function we investigated was the manual motor speed, and that we assessed it only with one test. Although finger tapping is regarded a reliable measure of the motor speed [67,68], a more extended evaluation is needed to verify our findings and investigate the effects of PBDE's on other aspects of the motor function than the manual motor speed.

Although we corrected for a number of potential confounders, we can not exclude that the associations we observed resulted from some source of confounding we failed to account for. Furthermore, a cross-sectional nature of our study does not allow to draw causal conclusions. We can not exclude the possibility that children with poor motor capabilities chose activities involving BFR exposure, which resulted in higher blood levels of these toxicants. Similarly, we can not be sure that the BFR exposure was a causing factor in the association between the toxicants and the thyroid function we observed. Our study had a fairly low response-rate. However, the comparison of socioeconomical status indicators between a subgroup of participants and non-participants did not reveal evidence of a selection bias.

13.5 CONCLUSIONS

Our study is one of few studies and so far the largest one investigating the neurobehavioral effects of BFR in humans. Low-level PBDE exposure was associated with changes in the motor function and serum levels of FT3 and TSH. Our observations need to be further elucidated in other age groups preferably using prospectively designed studies.

REFERENCES

1. De Wit C: An overview of brominated flame retardants in the environment. Chemosphere 2002, 46:583-624.
2. Birnbaum L, Staskal D: Brominated flame retardants: cause for concern? Environ Health Perspect 2004, 112:9-17.
3. Darnerud P, Eriksen G, Jóhannesson T, Larsen P, Viluksela M: Polybrominated diphenyl ethers: occurrence, dietary exposure, and toxicology. Environ Health Perspect 2001, 109:49-68.
4. Law RJ, Alaee M, Allchin CR, Boon JP, Lebeuf M, Lepom P, Stern GA: Levels and trends of polybrominated diphenylethers and other brominated flame retardants in wildlife. Environ Int 2003, 29:757-770.
5. Watanabe I, Sakai S: Environmental release and behavior of brominated flame retardants. Environ Int 2003, 29:665-682.
6. Roosens L, D'Hollander W, Bervoets L, Reynders H, Van Campenhout K, Cornelis C, Van Den Heuvel R, Koppen G, Covaci A: Brominated flame retardants and perfluorinated chemicals, two groups of persistent contaminants in Belgian human blood and milk. Environ Pollut 2010, 158:2546-2552.
7. Wu N, Herrmann T, Paepke O, Tickner J, Harvey L, La Guardia M, McClean M, Webster T: Human exposure to PBDEs: associations of PBDE body burdens with food consumption and house dust concentrations. Environ Sci Technol 2007, 41:1584-1589.
8. Costa L, Giordano G, Tagliaferri S, Caglieri A, Mutti A: Polybrominated diphenyl ether (PBDE) flame retardants: environmental contamination, human body burden and potential adverse health effects. Acta Biomed 2008, 79:172-183.
9. Dingemans M, van den Berg M, Westerink R: Neurotoxicity of brominated flame retardants: (in)direct effects of parent and hydroxylated polybrominated diphenyl

ethers on the (developing) nervous system. Environ Health Perspect 2011, 119:900-907.

10. Williams AL, DeSesso JM: The potential of selected brominated flame retardants to affect neurological development. J Toxicol Environ, Part B 2010, 13:411-448.

11. Eriksson P, Jakobsson E, Fredriksson A: Brominated flame retardants: a novel class of developmental neurotoxicants in our environment? Environ Health Perspect 2001, 109:903-908.

12. Gee JR, Moser VC: Acute postnatal exposure to brominated diphenylether 47 delays neuromotor ontogeny and alters motor activity in mice. Neurotoxicol Teratol 2008, 30:79-87.

13. Branchi I, Capone F, Vitalone A, Madia F, Santucci D, Alleva E, Costa LG: Early developmental exposure to BDE 99 or Aroclor 1254 affects neurobehavioural profile: interference from the administration route. Neurotoxicology 2005, 26:183-192.

14. Eriksson P, Viberg H, Jakobsson E, Orn U, Fredriksson A: A brominated flame retardant, 2,2',4,4',5-Pentabromodiphenyl ether: uptake, retention, and induction of neurobehavioral alterations in mice during a critical phase of neonatal brain development. Toxicol Sci 2002, 67:98-103.

15. Eriksson P, Fischer C, Fredriksson A: Polybrominated diphenyl ethers, a group of brominated flame retardants, can interact with polychlorinated biphenyls in enhancing developmental neurobehavioral defects. Toxicol Sci 2006, 94:302-309.

16. Viberg H, Fredriksson A, Eriksson P: Investigations of strain and/or gender differences in developmental neurotoxic effects of polybrominated diphenyl ethers in mice. Toxicol Sci 2004, 81:344-353.

17. Viberg H, Fredriksson A, Eriksson P: Neonatal exposure to polybrominated diphenyl ether (PBDE 153) disrupts spontaneous behaviour, impairs learning and memory, and decreases hippocampal cholinergic receptors in adult mice. Toxicol Appl Pharmacol 2003, 192:95-106.

18. Viberg H, Johansson N, Fredriksson A, Eriksson J, Marsh G, Eriksson P: Neonatal exposure to higher brominated diphenyl ethers, hepta-, octa-, or nonabromodiphenyl ether, impairs spontaneous behavior and learning and memory functions of adult mice. Toxicol Sci 2006, 92:211-218.

19. Johansson N, Viberg H, Fredriksson A, Eriksson P: Neonatal exposure to decabrominated diphenyl ether (PBDE 209) causes dose response changes in spontaneous behaviour and cholinergic susceptibility in adult mice. Neurotoxicology 2008, 29:911-919.

20. Viberg H, Fredriksson A, Eriksson P: Changes in spontaneous behaviour and altered response to nicotine in the adult rat, after neonatal exposure to the brominated flame retardant, decabrominated diphenyl ether (PBDE 209). Neurotoxicology 2007, 28:136-142.

21. Viberg H, Fredriksson A, Jakobsson E, Orn U, Eriksson P: Neurobehavioral derangements in adult mice receiving decabrominated diphenyl ether (PBDE 209) during a defined period of neonatal brain development. Toxicol Sci 2003, 76:112-120.

22. Rice DC, Reeve EA, Herlihy A, Thomas Zoeller R, Douglas Thompson W, Markowski VP: Developmental delays and locomotor activity in the C57BL6/J mouse following neonatal exposure to the fully-brominated PBDE, decabromodiphenyl ether. Neurotoxicol Teratol 2007, 29:511-520.

23. Dufault C, Poles G, Driscoll LL: Brief postnatal PBDE exposure alters learning and the cholinergic modulation of attention in rats. Toxicol Sci 2005, 88:172-180.

24. Driscoll LL, Gibson AM, Hieb A: Chronic postnatal DE-71 exposure: effects on learning, attention and thyroxine levels. Neurotoxicol Teratol 2009, 31:76-84.

25. Eriksson P, Viberg H, Fischer M, Wallin M, Fredriksson A: A comparison on developmental neurotoxic effects of hexabromocyclododecane, 2,2'4,4',5,5'-hexabromodiphenyl ether. Organohalogen Compd 2002, 57:389-392.

26. Eriksson P, Fischer C, Wallin M, Jakobsson E, Fredriksson A: Impaired behaviour, learning and memory, in adult mice neonatally exposed to hexabromocyclododecane (HBCDD). Environ Toxicol Pharmacol 2006, 21:317-322.

27. Ta TA, Koenig CM, Golub MS, Pessah IN, Qi L, Aronov PA, Berman RF: Bioaccumulation and behavioral effects of 2,2',4,4'-tetrabromodiphenyl ether (BDE-47) in perinatally exposed mice. Neurotoxicol Teratol 2011, 33:393-404.

28. Suvorov A, Girard S, Lachapelle S, Abdelouahab N, Sebire L, Guillaume S, Takser L: Perinatal exposure to low-dose BDE-47, an emergent environmental contaminant, causes hyperactivity in rat offspring. Neonatology 2009, 95:203-209.

29. Branchi I, Alleva E, Costa LG: Effects of perinatal exposure to a Polybrominated Diphenyl Ether (PBDE 99) on mouse neurobehavioural development. Neurotoxicology 2002, 23:375-384.

30. Kuriyama S, Talsness C, Grote K, Chahoud I: Developmental exposure to low-dose PBDE-99: Effects on Male Fertility and Neurobehavior in Rat Offspring. Environ Health Perspect 2005, 113:149-154.

31. Chou CT, Hsiao YC, Ko FC, Cheng JO, Cheng YM, Chen TH: Chronic exposure of 2,2',4,4'-tetrabromodiphenyl ether (PBDE-47) alters locomotion behavior in juvenile zebrafish (Danio rerio). Aquat Toxicol 2010, 98:388-395.

32. McClain V, Stapleton HM, Tilton F, Gallagher EP: BDE 49 and developmental toxicity in zebrafish. Comp Biochem Physiol C 2012, 155:253-258.

33. He P, He W, Wang A, Xia T, Xu B, Zhang M, Chen X: PBDE-47-induced oxidative stress, DNA damage and apoptosis in primary cultured rat hippocampal neurons. Neurotoxicology 2008, 29:124-129.

34. Huang SC, Giordano G, Costa LG: Comparative cytotoxicity and intracellular accumulation of five Polybrominated Diphenyl Ether congeners in mouse cerebellar granule neurons. Toxicol Sci 2010, 114:124-132.

35. Tagliaferri S, Caglieri A, Goldoni M, Pinelli S, Alinovi R, Poli D, Pellacani C, Giordano G, Mutti A, Costa LG: Low concentrations of the brominated flame retardants BDE-47 and BDE-99 induce synergistic oxidative stress-mediated neurotoxicity in human neuroblastoma cells. Toxicology in Vitro 2010, 24:116-122.

36. Chao H, Tsou T, Huang H, Chang-Chien G: Levels of breast milk PBDEs from southern Taiwan and their potential impact on neurodevelopment. Pediatr Res 2011, 70:596-600.

37. Herbstman JB, Sjodin A, Kurzon M, Lederman SA, Jones RS, Rauh V, Needham LL, Tang D, Niedzwiecki M, Wang RY, Perera F: Prenatal Exposure to PBDEs and Neurodevelopment. Environ Health Perspect 2010, 118:712-719.

38. Roze E, Meijer L, Bakker A, Van Braeckel KNJA, Sauer PJJ, Bos AF: Prenatal exposure to organohalogens, including brominated flame retardants, influences motor,

cognitive, and behavioral performance at school age. Environ Health Perspect 2009, 117:1953-1958.

39. Fitzgerald EF, Shrestha S, Gomez MI, McCaffrey RJ, Zimmerman EA, Kannan K, Hwang S: Polybrominated diphenyl ethers (PBDEs), polychlorinated biphenyls (PCBs) and neuropsychological status among older adults in New York. Neurotoxicology 2012, 33:8-15.

40. Legler J: New insights into the endocrine disrupting effects of brominated flame retardants. Chemosphere 2008, 73:216-222.

41. Darnerud PO, Aune M, Larsson L, Hallgren S: Plasma PBDE and thyroxine levels in rats exposed to Bromkal or BDE-47. Chemosphere 2007, 67:S386-S392.

42. Richardson VM, Staskal DF, Ross DG, Diliberto JJ, DeVito MJ, Birnbaum LS: Possible mechanisms of thyroid hormone disruption in mice by BDE 47, a major polybrominated diphenyl ether congener. Toxicol Appl Pharmacol 2008, 226:244-250.

43. Stoker TE, Laws SC, Crofton KM, Hedge JM, Ferrell JM, Cooper RL: Assessment of DE-71, a commercial Polybrominated Diphenyl Ether (PBDE) mixture, in the EDSP male and female pubertal protocols. Toxicol Sci 2004, 78:144-155.

44. Zhou T, Ross DG, DeVito MJ, Crofton KM: Effects of short-term in vivo exposure to Polybrominated Diphenyl Ethers on thyroid hormones and hepatic enzyme activities in weanling rats. Toxicol Sci 2001, 61:76-82.

45. Zhou T, Taylor MM, DeVito MJ, Crofton KM: Developmental exposure to brominated diphenyl ethers results in thyroid hormone disruption. Toxicol Sci 2002, 66:105-116.

46. Abdelouahab N, Suvorov A, Pasquier J, Langlois M, Praud J, Takser L: Thyroid disruption by low-dose BDE-47 in prenatally exposed lambs. Neonatology 2009, 96:120-124.

47. Lee E, Kim T, Choi J, Nabanata P, Kim N, Ahn M, Jung K, Kang I, Kim T, Kwack S, Park KL, Kim SH, Kang TS, Lee J, Lee BM, Kim HS: Evaluation of liver and thyroid toxicity in Sprague–Dawley rats after exposure to polybrominated diphenyl ether BDE-209. J Toxicol Sci 2010, 35:535-545.

48. Zhang S, Bursian SJ, Martin PA, Chan HM, Tomy G, Palace VP, Mayne GJ, Martin JW: Reproductive and developmental toxicity of a pentabrominated diphenyl ether mixture, DE-71, to ranch mink (Mustela vison) and hazard assessment for wild mink in the Great Lakes Region. Toxicol Sci 2009, 110:107-116.

49. Fonnum F, Mariussen E: Mechanisms involved in the neurotoxic effects of environmental toxicants such as polychlorinated biphenyls and brominated flame retardants. J Neurochem 2009, 111:1327-1347.

50. Baker EL, Letz RE, Fidler AT, Shalat S, Plantamura D, Lyndon M: A computer-based neurobehavioral evaluation system for occupational and environmental epidemiology: methodology and validation studies. Neurobehav Toxicol Teratol 1985, 7:369-377.

51. White RF, James KE, Vasterling JJ, Letz R, Marans K, Delaney R, Krengel M, Rose F, Kraemer HC: Neuropsychological screening for cognitive impairment using computer-assisted tasks. Assessment 2003, 10:86-101.

52. Letz R: NES3 user's manual. Neurobehavioral Systems Inc., Atlanta (GA); 2000.

53. Covaci A, Voorspoels S: Optimization of the determination of polybrominated diphenyl ethers in human serum using solid-phase extraction and gas chromatogra-

phy-electron capture negative ionization mass spectrometry. J Chromatogr B 2005, 827:216-223.

54. Schroijen C, Baeyens W, Schoeters G, Den Hond E, Koppen G, Bruckers L, Nelen V, Van De Mieroop E, Bilau M, Covaci A, Keune H, Loots I, Kleinjans J, Dhooge W, Van Larebeke N: Internal exposure to pollutants measured in blood and urine of Flemish adolescents in function of area of residence. Chemosphere 2008, 71:1317-1325.

55. Lin SM, Chen FA, Huang YF, Hsing LL, Chen LL, Wu LS, Liu TS, Chang-Chien GP, Chen KC, Chao HR: Negative associations between PBDE levels and thyroid hormones in cord blood. Int J Hyg Environ Health 2011, 214:115-120.

56. Turyk M, Persky V, Imm P, Knobeloch L, Chatterton R, Anderson H: Hormone disruption by PBDEs in adult male sport fish consumers. Environ Health Perspect 2008, 116:1635-1641.

57. Chevrier J, Harley K, Bradman A, Gharbi M, Sjödin A, Eskenazi B: Polybrominated diphenyl ether (PBDE) flame retardants and thyroid hormone during pregnancy. Environ Health Perspect 2010, 118:1444-1449.

58. Herbstman J, Sjödin A, Apelberg B, Witter F, Halden R, Patterson D, Panny S, Needham L, Goldman L: Birth delivery mode modifies the associations between prenatal polychlorinated biphenyl (PCB) and polybrominated diphenyl ether (PBDE) and neonatal thyroid hormone levels. Environ Health Perspect 2008, 116:1376-1382.

59. Stapleton HM, Eagle S, Anthopolos R, Wolkin A, Miranda ML: Associations between Polybrominated Diphenyl Ether (PBDE) flame retardants, phenolic metabolites, and thyroid hormones during pregnancy. Environ Health Perspect 2011, 119:1454-1459.

60. Talsness CE, Kuriyama SN, Sterner-Kock A, Schnitker P, Grande S, Shakibaei M, Andrade A, Grote K, Chahoud I: In Utero and Lactational Exposures to Low Doses of Polybrominated Diphenyl Ether-47 Alter the Reproductive System and Thyroid Gland of Female Rat Offspring. Environ Health Perspect 2007, 116:308-314.

61. Viberg H, Fredriksson A, Eriksson P: Neonatal exposure to the brominated flame retardant 2,2',4,4',5-pentabromodiphenyl ether causes altered susceptibility in the cholinergic transmitter system in the adult mouse. Toxicol Sci 2002, 67:104-107.

62. Viberg H, Fredriksson A, Eriksson P: Neonatal exposure to the brominated flame-retardant, 2,2,4,4,5-pentabromodiphenyl ether, decreases cholinergic nicotinic receptors in hippocampus and affects spontaneous behaviour in the adult mouse. Environ Toxicol Pharmacol 2004, 17:61-65.

63. Geyer HJ, Schramm K-W, Darnerud PO, Aune M, Feicht EA, Fried KW, Henkelmann B, Lenoir D, Schmid P, McDonald TA: Terminal elimination half-lives of the brominated flame retardants TBBPA, HBCD, and lower brominated PBDES in humans. Organohalogen Compd 2004, 66:3820-3825.

64. Thuresson K, Höglund P, Hagmar L, Sjödin A, Bergman A, Jakobsson K: Apparent half-lives of hepta- to decabrominated diphenyl ethers in human serum as determined in occupationally exposed workers. Environ Health Perspect 2006, 114:176-181.

65. Trudel D, Scheringer M, Von Goetz N, Hungerbuhler K: Total consumer exposure to polybrominated diphenyl ethers in North America and Europe. Environ Sci Technol 2011, 45:2391-2397.

66. Szymanska J, Sapota A, Frydrych B: The disposition and metabolism of tetrabromobisphenol-A after a single i.p. dose in the rat. Chemosphere 2001, 45:693-700.
67. Mitrushina M, Boone K, Razani J, D'Elia L: Handbook of Normative Data for Neuropsychological Assessment. Oxford University Press, Oxford; 2005.
68. Lezak M, Howieson D, Loring D: Neuropsychological Assessment. Oxford University Press, Oxford; 2004.

There is one table that is not available in this version of the article. To view this additional information, please use the citation on the first page of this chapter.

PART VII

LOOKING TOWARD THE FUTURE

CHAPTER 14

UNCERTAIN INHERITANCE: TRANSGENERATIONAL EFFECTS OF ENVIRONMENTAL EXPOSURES

CHARLES W. SCHMIDT

Andrea Cupp made a serendipitous discovery when she was a postdoctoral fellow at Washington State University: While investigating how chemicals affect sex determination in embryonic animals, she bred the offspring of pregnant rats that had been dosed with an insecticide called methoxyclor. When the males from that litter grew into adults, they had decreased sperm counts and higher rates of infertility. Cupp had seen these same abnormalities in the animals' fathers, which had been exposed to methoxyclor in the womb. But this latest generation hadn't been exposed that way, which suggested that methoxyclor's toxic effects had carried over generations. "At first I couldn't believe it," says Cupp's advisor, Michael Skinner, a biochemist and Washington State professor. "But then we repeated the breeding experiments and found that the results held up."

Skinner and Cupp, who is now a professor at the University of Nebraska–Lincoln, published their findings in 2005. [1] Since that paper—which

Used with permission from Environmental Health Perspectives. Schmidt CW. Uncertain Inheritance: Transgenerational Effects of Environmental Exposures. Environmental Health Perspectives, 121 (2013), DOI:10.1289/ehp.121-A298.

showed that reproductive effects not just from methoxyclor but also from the fungicide vinclozolin persisted for at least four generations—the number of published articles reporting similar transgenerational findings has increased steadily. "In the last year and half there's been an explosion in studies showing transgenerational effects from exposure to a wide array of environmental stressors," says Lisa Chadwick, a program administrator at the National Institute of Environmental Health Sciences (NIEHS). "This is a field that's really starting to take off."

According to Chadwick, the new findings compel a reevaluation of how scientists perceive environmental health threats. "We have to think more long-term about the effects of chemicals that we're exposed to every day," she says. "This new research suggests they could have consequences not just for our own health and for that of our children, but also for the health of generations to come."

The NIEHS recently issued requests for applications totaling $3 million for research on transgenerational effects in mammals. [2] Chadwick says funded studies will address two fundamental data needs, one pertaining to potential transgenerational mechanisms and another to the number of chemicals thought to exert these effects. These studies will extend to what's known as the F3 generation—the great-grandchildren of the originally exposed animal. That's because chemicals given to pregnant females (the F0 generation) interact not only with the fetal offspring (the F1 generation) but also the germ cells developing within those offspring, which mature into the sperm and eggs that give rise to the F2 generation. Thus, the F3 animals are the first generation to be totally unexposed to the original agent. Effects that extend to the F2 generation are known as "multigenerational," whereas those that extend to the F3 generation are known as "transgenerational." [3]

Transgenerational effects have now been reported for chemicals including permethrin, DEET, bisphenol A, certain phthalates, dioxin, jet fuel mixtures, nicotine, and tributyltin, among others. Most of these findings come from rodent studies. [4,5,6,7] But preliminary evidence that chemical effects can carry over generations in humans is also emerging, although no F3 data have been published yet. Given the challenges of tracking effects over multiple human lifespans, the evidence is more difficult to interpret, particularly with respect to potential mechanisms, says Tessa

Roseboom, a professor of early development and health at the Academic Medical Center in Amsterdam, the Netherlands. Still, some reports have linked nutritional deficiencies from famine and exposure to diethylstilbestrol (DES)—a nonsteroidal estrogen used to protect against miscarriage from the 1940s to the 1970s—to effects that persist among the grandchildren of exposed women. [8,9,10,11,12,13]

14.1 FOUNDATIONS IN ANIMAL DATA

The way in which environmental exposures cause transgenerational effects is unclear. According to Chadwick, current hypotheses lean toward epigenetic inheritance patterns, which involve chemical modifications to the DNA rather than mutations of the DNA sequence itself. Scientists debate the precise definition of "epigenetics," but Robert Waterland, an associate professor of pediatrics and genetics at Baylor College of Medicine, suggests the best definition was published in Nature Genetics 10 years ago: "The study of stable alterations in gene expression potential that arise during development and cell proliferation." [14]

Epigenetic modifications can take a few different forms—molecules known as methyl groups can attach to DNA itself, or methyl or acetyl groups can attach to the histone proteins that surround DNA. These attached molecules, also known as "marks" or "tags," influence gene expression and thereby determine the specialized function of every cell in the organism.

Epigenetic marks carried over from the parents are typically wiped clean during molecular programming events that happen early in embryonic development. Shortly after fertilization, explains Dana Dolinoy, an assistant professor at the University of Michigan School of Public Health, a wave of DNA demethylation leaves the embryo with a fresh genomic slate with the exception of certain imprinted genes, such as insulin-like growth factor 2 (IGF2), which remain methylated. Later, cells in the developing embryo are remethylated as they develop into the somatic cells that make up different organs and tissues in the body. Germ cells, meanwhile, undergo their own wave of demethylation and remethylation programming events, which are specific to the sex of the developing embryo.

Researchers have found that transgenerational effects can result from chemical dosing at precise windows in fetal development—specifically, at the time of sex determination, which occurs around embryonic days 10.5–12.5 for mice and embryonic days 41–44 for humans, according to Duke University cell biology professor Blanche Capel. Observable effects in the F3 generation are thought to result from changes to the germ line, which is the succession of germ cell DNA that passes from one generation to the next. Skinner and other researchers have identified DNA methylation changes in F3-generation sperm that appear to underlie transgenerational effects seen in F3 animals. [1,4]

Researchers emphasize that much of the evidence so far in the field is observational, meaning the biological mechanisms remain unknown. Dolinoy says scientific opinions lean heavily toward epigenetic pathways. "That seems to be where the whole field is headed," she says.

According to Chadwick, Skinner's laboratory remains a nexus for transgenerational studies in chemically exposed animals. In his more recent work, Skinner has shown that insecticides, phthalates, dioxin, and jet fuel, when given to gestating rats during periods of embryonic programming, promote early-onset puberty in female offspring and decreased sperm counts in males, out to the F3 generation. [4] "We mapped DNA methylation in germ cells and found that each compound induces a unique epigenetic signature," Skinner says. "But it's also possible that other epigenetic mechanisms play a role."

Meanwhile, several other groups are studying transgenerational changes in animals. In one study, Kwan Hee Kim, a professor of molecular biosciences at Washington State, exposed pregnant mice to di-(2-ethylhexyl) phthalate (DEHP) on embryonic days 7–14. [7] Kim observed decreased sperm counts and sperm motility in male offspring out to the F4 generation. Importantly, she also observed an 80% reduction in spermatogonial stem cell regeneration. Consequently, she says, "As the animals aged, their ability to make new sperm decreased dramatically."

Kim implicates DNA methylation as a potential epigenetic mechanism behind the change in function. During the study, she identified 16 genes that were differentially methylated and expressed in newborn pups, she says. This group of targeted genes may hold clues to how DEHP acts transgenerationally.

In another new study, Virender Rehan, a professor of pediatrics at the Harbor–UCLA Medical Center, found that prenatal exposure to nicotine in rats starting at embryonic day 6 was associated with asthma-like symptoms among F3 males and females. But, similarly to an earlier study extending to F2 offspring, the effects were sex-specific, with total airway system resistance significantly greater in males than females, due in part to tracheal constriction, which was detected only in males. [5,6] What's still unclear (and a subject of his current research), Rehan says, is whether the transgenerational effect is being carried through the male or female germ line.

Bruce Blumberg, a professor of developmental and cell biology at the University of California, Irvine, recently published a mouse study showing that maternal exposure to the biocide tributyltin (TBT) induced a condition similar to nonalcoholic fatty liver disease out to the F3 generation. [15] Like other transgenerational toxicants, TBT is an endocrine disruptor that appears to be an obesogen, or a chemical that promotes obesity partly by promoting the growth of fat cells. [16] Blumberg's study used doses as much as 50-fold lower than the no observed adverse effect level (NOAEL) for TBT.

According to Blumberg, the findings also support an evolving concept in reproductive biology—the "developmental origins of health and disease" hypothesis, which holds that low-dose chemical exposures or maternal dietary changes experienced in utero can induce permanent physical changes in adult animals. [17] "These effects are permanent in that they remain even when you take away the exposure," he says. "Now we're finding that the effects can also last through subsequent generations."

Other researchers have found evidence that transgenerational effects can impact mating behaviors, with implications for the evolution of populations. In one example, David Crews, a professor of biology and psychology at the University of Texas at Austin, reported that female rats avoided F3 males with an ancestral exposure to vinclozolin. The study specifically found that all females tested preferred control males (who had no ancestral vinclozolin exposure) whereas males from both the control and ancestrally treated groups exhibited no particular preference for female type. [18] "Where the rubber meets the road in evolution is sex," Crews says. "It's all about who mates and reproduces with who."

14.2 THE CASE FOR MULTIGENERATIONAL EFFECTS IN HUMANS

The evidence for environmentally induced multigenerational effects in humans began to emerge years ago from an isolated community in Northern Sweden called Överkalix Parish. Led in part by Marcus Pembrey, a clinical geneticist at the University College London Institute of Child Health, researchers investigated whether an abundance of food in childhood had any influence on the risk of heart disease and diabetes among a child's future descendants. In particular, the researchers studied overeating during a child's "slow-growth period," the lull before the prepubertal growth spurt.

An initial study published in 2002 suggested the answer was a conditional yes. By studying harvest statistics, grain prices, and other records, the researchers classified food availability in Överkalix for individual years of the nineteenth century as poor, moderate, or good. They then studied health outcomes among descendants born in 1890, 1905, and 1920, and found that food abundance during the grandfather's (but not grandmother's) slow-growth period was associated with an increase in diabetes mortality. [19]

In a follow-up study of the same Överkalix individuals, Pembrey and colleagues found further evidence of sex-specific multigenerational effects: Male descendants had a statistically increased relative risk of mortality if the paternal grandfather had a good food supply during his slow-growth period, while females had statistically higher relative risks if the paternal grandmother had good food availability during her slow-growth period. [12]

Other data come from the grandchildren of women who were pregnant in the Western Netherlands in the winter of 1944–1945, when nutritional intake dropped to as little as 400 calories per day as a result of food import restrictions by the occupying German army. In 2008 researchers led by Roseboom reported that the children of women who were exposed to famine in utero tended to be fatter at birth and more prone to health problems in adulthood than the children of women born before or after the famine. [10] In earlier studies, Roseboom and colleagues had reported that adult F1 populations exposed to famine conditions in utero had higher rates of cardiovascular disease, [20,21] diabetes, [22,23] obesity, [24] and breast cancer. [25]

The F1 mothers in the 2008 study completed questionnaires about the birth conditions and current health status of their grown children. The questionnaire grouped health outcomes into four categories: congenital, cardiovascular and metabolic, psychiatric, and other. The only statistically significant association between ancestral famine exposure and poor health outcomes was with the "other" category, which included accidents and acquired neurological, autoimmune, infectious, respiratory, neoplastic, and dermatological conditions. In their conclusions, Roseboom and colleagues state that the findings "constitute the first direct evidence in humans that the detrimental effects of poor maternal nutrition during gestation on health in later life pass down to subsequent generations." [10]

Roseboom calls the findings "a first but weak" indication of multigenerational effects on health after prenatal famine exposure. "It was weak because we approached the F1 and not the F2 directly," she explains. "But in a next study we contacted the F2 directly, and we found they were more adipose not only at birth but also currently while in their forties, and therefore we expect that they might have increased cardiovascular disease rates later on in their lives."

Another key line of human evidence in the field comes from multigenerational studies of DES. [9] Those data came from a pair of National Cancer Institute studies: the DES Follow-Up Study, which tracks health outcomes among women who were exposed to DES and their prenatally exposed children, and the DES Third Generation Cohort Study, which tracks the male and female grandchildren of the originally exposed women.

According to Linda Titus, a professor in community and family medicine and pediatrics at the Geisel School of Medicine at Dartmouth, grandsons of DES-exposed women had a modestly higher risk of any birth defect, mostly urogenital defects, although the findings weren't statistically significant. Granddaughters, meanwhile, had a higher frequency of hip dysplasia, irregular periods, older age at menarche, and potentially an increased risk of infertility. There was also a higher risk of ovarian cancer among granddaughters of exposed women, but since that finding is based on just three cases, she says, it must be considered preliminary. [26]

14.3 EPIGENETIC EVIDENCE IN HUMANS STILL EMERGING

The human data on epigenetics is generally limited to F1 populations and comes mainly from studies on the Dutch famine. [8,10,11,12] According to Roseboom, the first study to link undernutrition during gestation to altered epigenetic status was published by Bastian T. Heijmans, an associate professor of genetics at Leiden University Medical Center.8 In that study, Heijmans and colleagues reported that F1 generations exposed to Dutch famine conditions in utero had hypomethylation of the IGF2 gene six decades later, compared with same-sex siblings not exposed to famine (they noted that other stressors such as cold and emotional stress could have contributed to the observed hypomethylation).

According to Roseboom, this finding suggests prenatal famine could lead to changes in gene expression via changes in methylation. But Heijmans' research team was not able to statistically associate hypomethylated IGF2 with any specific health outcomes. And Roseboom points out that "whether these changes in methylation actually result in changes in gene expression and ultimately changes in, for instance, cardiovascular risk factors remains to be investigated."

Roseboom's team followed up last year with a study investigating four additional genes that have been shown in animals to be persistently altered by maternal dietary restrictions. But the study failed to demonstrate any consistent links between famine exposure and methylation status, possibly because of confounding from lifestyle choices and diet later in life. [13] Roseboom's team is currently analyzing methylation levels on DNA obtained from the F0, F1, and F2 generations affected by the Dutch famine; these data have not yet been submitted for publication.

Titus says that conclusive evidence of transgenerational epigenetic mechanisms in humans will depend on findings in F3 generations. "Even if new studies confirm outcomes in DES-exposed grandchildren, we can't be sure if they are due to epigenetic changes," she says. "A true assessment of heritable epigenetic changes requires studies of great-grandchildren, which will be the first generation without DES exposure."

Blumberg emphasizes that just because the data haven't yet materialized doesn't mean that environmentally induced, transgenerational epigenetic changes in humans don't occur. "We see transgenerational epigenetic changes in animals, and what we believe is that the animal data predict human responses," he says. "Moreover, it's possible that you won't see epigenetic changes from looking at genes—you might see it, instead, in noncoding regions in DNA."

The growing evidence that environmental exposures might induce a myriad of effects that persist transgenerationally leaves open questions about where human evolution is headed, Crews asserts. "It's a new window on the 'nature versus nurture' debate," he says. "We're all combinations of what we inherit and what we're exposed to in our own lives. And right now you can't find a human or an animal on the planet without a body burden of endocrine-disrupting chemicals."

REFERENCES AND NOTES

1. Anway MD, et al. Epigenetic transgenerational actions of endocrine disruptors and male fertility. Science 308(5727):1466–1469 (2005); http://dx.doi.org/10.1126/science.1108190.
2. Other National Institutes of Health entities, including the National Institute on Alcohol Abuse and Alcoholism and the National Institute on Drug Abuse, have launched parallel programs.
3. Skinner MK. What is an epigenetic transgenerational phenotype? F3 or F2. Reprod Toxicol 25(1):2–6 (2008); http://dx.doi.org/10.1016/j.reprotox.2007.09.001.
4. Manikkam M, et al. Transgenerational actions of environmental compounds on reproductive disease and identification of epigenetic biomarkers of ancestral exposures. PLoS ONE 7(2):e31901 (2012); http://dx.doi.org/10.1371/journal.pone.0031901.
5. Rehan VK, et al. Perinatal nicotine exposure induces asthma in second generation offspring. BMC Med 10:129 (2012); http://dx.doi.org/10.1186/1741-7015-10-129.
6. Rehan VK. Perinatal nicotine-induced transgenerational asthma. Am J Physiol Lung Cell Mol Physiol; http://dx.doi.org/10.1152/ajplung.00078.2013 [online 2 August 2013].
7. Doyle TJ, et al. Transgenerational effects of di-(2-ethylhexyl) phthalate on testicular germ cell associations and spermatogonial stem cells in mice. Biol Reprod 88(5):112 (2013); http://dx.doi.org/10.1095/biolreprod.112.106104.
8. Heijmans BT, et al. Persistent epigenetic differences associated with prenatal exposure to famine in humans. Proc Natl Acad Sci USA 105(44):17046–17049 (2008); http://dx.doi.org/10.1073/pnas.0806560105.
9. Titus-Ernstoff L, et al. Birth defects in the sons and daughters of women who were exposed in utero to diethylstilbestrol (DES). Int J Androl 33(2):377–384 (2010); http://dx.doi.org/10.1111/j.1365-2605.2009.01010.x.

10. Painter RC, et al. Transgenerational effects of prenatal exposure to the Dutch famine on neonatal adiposity and health in later life. BJOG 115(10):1243–1249 (2008); http://dx.doi.org/10.1111/j.1471-0528.2008.01822.x.

11. Veenendaal MV, et al. Transgenerational effects of prenatal exposure to the 1944–45 Dutch famine. BJOG 120(5):548–553 (2013); http://dx.doi.org/10.1111/1471-0528.12136.

12. Pembrey ME, et al. Sex-specific, male-line transgenerational responses in humans. Eur J Hum Genet 14(2):159–166 (2006); http://dx.doi.org/10.1038/sj.ejhg.5201538.

13. Veenendaal MV, et al. Prenatal famine exposure, health in later life and promoter methylation of four candidate genes. J Develop Orig Health Dis 3(8):450–457 (2012); http://dx.doi.org/10.1017/S2040174412000396.

14. Jaenisch R, Bird A.. Epigenetic regulation of gene expression: how the genome integrates intrinsic and environmental signals. Nat Genet 33(3s):245–254 (2003); http://dx.doi.org/10.1038/ng1089.

15. Chamorro-Garcia R, et al. Transgenerational inheritance of increased fat depot size, stem cell reprogramming, and hepatic steatosis elicited by prenatal exposure to the obesogen tributyltin in mice. Environ Health Perspect 121(3):359–366 (2013); http://dx.doi.org/10.1289/ehp.1205701.

16. Grün F, et al. Endocrine-disrupting organotin compounds are potent inducers of adipogenesis in vertebrates. Mol Endocrinol 20(9):2141–2155 (2006); http://dx.doi.org/10.1210/me.2005-0367.

17. Gluckman PD, Hanson MA. Developmental origins of disease paradigm: a mechanistic and evolutionary perspective. Pediatr Res 56(3):311–317 (2004); http://dx.doi.org/10.1203/01.PDR.0000135998.08025.FB.

18. Crews D, et al. Transgenerational epigenetic imprints on mate preference. Proc Natl Acad Sci USA 104(14):5942–5946 (2007); http://dx.doi.org/10.1073/pnas.0610410104.

19. Kaati G, et al. Cardiovascular and diabetes mortality determined by nutrition during parents' and grandparents' slow growth period. Eur J Hum Genet 10(11):682–688 (2002); http://dx.doi.org/10.1038/sj.ejhg.5200859.

20. Painter RC, et al. Early onset of coronary artery disease after prenatal exposure to the Dutch famine. Am J Clin Nutr 84(2):322–327 (2006); http://ajcn.nutrition.org/content/84/2/322.

21. Roseboom TJ, et al. Coronary heart disease after prenatal exposure to the Dutch famine, 1944–45. Heart 84:595–8 (2000); http://dx.doi.org/10.1136/heart.84.6.595.

22. de Rooij SR, et al. Glucose tolerance at age 58 and the decline of glucose tolerance in comparison with age 50 in people prenatally exposed to the Dutch famine. Diabetologia 49(4):637–643 (2006); http://link.springer.com/article/10.1007%2Fs00125-005-0136-9.

23. Ravelli ACJ, et al. Glucose tolerance in adults after prenatal exposure to famine. Lancet 351(9097):173–177 (1998); http://www.sciencedirect.com/science/article/pii/S0140673697072449.

24. Ravelli ACJ, et al. Obesity at the age of 50 y in men and women exposed to famine prenatally. Am J Clin Nutr 70(5):811–816 (1999); http://ajcn.nutrition.org/content/70/5/811.full.pdf+html.

25. Painter RC, et al. A possible link between prenatal exposure to famine and breast cancer: a preliminary study. Am J Hum Biol 18(6):853–856 (2006); http://onlinelibrary.wiley.com/doi/10.1002/ajhb.20564/abstract.

26. Titus-Ernstoff L, et al. Offspring of women exposed in utero to diethylstilbestrol (DES): a preliminary report of benign and malignant pathology in the third generation. Epidemiology 19(2):251–257 (2008); http://dx.doi.org/10.1097/EDE.0b013e318163152a.

CHAPTER 15

TAKING ACTION TO PROTECT CHILDREN FROM ENVIRONMENTAL HAZARDS (EXCERPT FROM *CHILDREN'S HEALTH AND THE ENVIRONMENT: A GLOBAL PERSPECTIVE*)

S. BOESE-O'REILLY AND M. K. E. SHIMKIN

In protecting children's environmental health, every level has a role to play, from members of the family and community to local, regional, national and international bodies. Everyone has a part in offering children the best chances in life, and in making a difference in how they live, grow, play, learn, develop and eventually work and become productive members of society. While noteworthy accomplishments at all levels reach a variety of people, much remains to be done to sustain progress and intenSify change. Certain countries have leapt into action while others hardly know of the concern. Global movements will narrow gaps between countries in the level of effort and involve progressively more regions of the world,

Reprinted with Permission from the WHO. Boese-O'Reilly S and Shimkin MKE. "Taking Action to Protect Children from Environmental Hazards." In Children's Health and the Environment: A Global Perspective, *edited by J. Pronczuk-Garbino, Geneva: World Health Organization, 2005.*

enhancing opportunities for children in all countries to have healthy and productive lives.

15.1 FAMILIES, CARERS AND TEACHERS

Parents, child care providers and teachers can make a tremendous difference in the health of children through actions at home, in the child care setting and at school. These adults can provide role models for healthy behaviour and teach and guide children to create healthy environments. Efforts to motivate teachers in Chinese schools to refrain from smoking, for example, have improved the quality of indoor air in the schools and may influence children not to smoke. "Tools for Schools" is a programme in the USA that teaches children, teachers and administrative staff to conduct indoor air audits in schools and take action to remedy sources of pollution. In Australia, many schools are joining the SunSmart Schools Program, which involves the whole school community in protecting children from overexposure to the sun. Parents, child care providers and teachers can find success acting alone or with others to improve children's environmental health. Individual and local level effort will make a difference for children and may have greater impact than anticipated.

15.2 IDEAS FOR ACTION AT THE LOCAL LEVEL

The United States Environmental Protection Agency has developed "Tips to protect children from environmental risks" (1), which have been disseminated through doctor's offices, schools and on the Internet. These are practical, action-oriented steps that parents and carers can take to protect children. Any individual or group could develop similar tips and share them as part of a community education effort, focusing on local issues of greatest concern for children's well-being. WHO has put together messages for creating healthy environments for children as shown in Fig. 20.1 (2).

15.2.1 TIPS TO PROTECT CHILDREN FROM ENVIRONMENTAL RISKS (DEVELOPED BY THE ENVIRONMENTAL PROTECTION AGENCY)

Help children breathe easier
- Don't smoke and don't let others smoke in your home or car.
- Keep your home as clean as possible. Dust, mould, certain household pests, secondhand smoke, and pet dander can trigger asthma attacks and allergies.
- Limit outdoor activity on "ozone alert" days when air pollution is especially harmful.
- Walk, use bicycles, join or form carpools, and take public transportation.
- Limit motor vehicle idling.
- Avoid open burning.

Protect children from lead poisoning
- Get kids tested for lead by their doctor or health care provider.
- Test your home for lead paint hazards if it was built before 1978.
- Wash children's hands before they eat; wash bottles, pacifiers, and toys often.
- Wash floors and window sills to protect kids from dust and peeling paint contaminated with lead—especially in older homes.
- Run the cold water for at least 30 seconds to flush lead from pipes.

Keep pesticides and other toxic chemicals away from children
- Store food and trash in closed containers to keep pests from coming into your home.
- Use baits and traps when you can; place baits and traps where kids can't get them.
- Read product labels and follow directions.
- Store pesticides and toxic chemicals where kids can't reach them—never put them in other containers that kids can mistake for food or drink.
- Keep children, toys, and pets away when pesticides are applied; don't let them play in fields, orchards, and gardens after pesticides have been used for at least the time recommended on the pesticide label.
- Wash fruits and vegetables under running water before eating-peel them before eating, when possible.

Protect children from carbon monoxide (CO) poisoning
- Have fuel-burning appliances, furnace fluids, and chimneys checked once a year.

- Never use gas ovens or burners for heat; never use barbecues or grills indoors or in the garage.
- Never sleep in rooms with un-vented gas or kerosene space heaters.
- Don't run cars or lawnmowers in the garage.
- Install a CO alarm that meets UL, lAS, or Canadian standards in sleeping areas.

Protect children from contaminated fish and polluted water
- Be alert for local fish advisories and beach closings. Contact your local health department.
- Take used motor oil to a recycling centre; properly dispose of toxic household chemicals.
- Learn what's in your drinking-water—call your local public water supplier for annual drinking water quality reports; for private drinking-water wells, have them tested annually by a certified laboratory. Call 1-800-426-4791 or contact www.epa.gov/safewater for help.

Safeguard children from high levels of radon
- Test your home for radon with a home test kit.
- Fix your home if your radon level is 4 pCi/L or higher. For help, call your state radon office or 1-aOO-SOS-RADON.

Protect children from too much sun
- Wear hats, sunglasses, and protective clothing.
- Use sunscreen with SPF 15+ on kids over six months; keep infants out of direct sunlight.
- Limit time in the midday sun-the sun is most intense between 10 and 4.

Keep children and mercury apart
- Eat a balanced diet but avoid fish with high levels of mercury.
- Replace mercury thermometers with digital thermometers.
- Don't let kids handle or play with mercury .
- Never heat or burn mercury.
- Contact your state or local health or environment department if mercury is spilled—never vacuum a spill.

15.2.2 COMMUNITIES AND LOCAL GOVERNMENTS

Enforcement of environmental regulations, housing initiatives, school and hospital administration, disease surveillance and reporting, and other public services usually fall within the jurisdiction of local governments, offering many opportunities to make communities "child friendly" in the environmental health context. Local governments often serve as first-line response to environmental incidents and are the key communicators with the general public. Not only are they responsible for communicating about risk and public safety, they also are the intermediaries between national policymakers and citizens. They can take proactive and preventive steps by promoting community events, sponsoring poison control centres, and creating innovative ways to educate and protect the public, especially children.

15.2.2.1 COMMUNITY MEETINGS LEAD TO REDUCTION IN CHILDHOOD ASTHMA

Particulate matter and harmful fumes are byproducts of the combustion process used in foundries that melt scrap iron. Residents of Sandwell, an urban area in the United Kingdom, became concerned about the respiratory health of their children, who attended schools close to such a foundry, and organized a public meeting to discuss the issue. The resulting measures taken by the foundry management led to a significant reduction in asthma-related hospital admissions of schoolchildren in the area (Case Study 19, page 342).

15.3 NATIONAL GOVERNMENTS

The words of a head of state can generate attention, political will and funding. National governments can gear environmental protection systems and structures to improve children's environmental health, supporting decentralized initiatives and working to formulate regulations and comply with national mandates. National governments are the ultimate champions for children's environmental health, and should monitor it through data collection and analysis. They should support communication and national public awareness efforts, pilot actions, specific projects and more. In addition, national governmments should contribute to international efforts that spread the message beyond their borders, promote collaboration and strengthen the agenda at home.

15.4 INTERNATIONAL AND GLOBAL EFFORTS

Over recent years, the international agenda has given considerable attention to children's environmental health, setting a framework for action by
individuals and entities worldwide. Recent examples of progress on the
international front include the following:

- At the World Summit on Sustainable Development in Johannesburg, South
  Africa, September 2002, the World Health Organization called for a global
  movement to improve children's environmental health, motivating countries, United Nations agencies and non-governmental organizations to create a "mass movement for children's environmental health." The organization has since taken steps to form a global Healthy Environments for
  Children Alliance to support countries as they strive to improve children's
  environmental health through national and local efforts (2). WHO programmes and regional offices have begun facilitating regional and national
  efforts to improve children's environmental health (5).
- The Commission for Environmental Cooperation of North America adopted
  an agenda for action on children's health and environment in June 2002 (4).
- In May 2002, the United Nations convened a General Assembly Special
  Session on children, which hosted side events on children's environmental
  health and resulted in a document that stressed the environment as an integral element of child health and welfare (5).
- Countries of South-East Asia and the Western Pacific were addressed by the
  International Conference on Children's Environmental Health: 260 Hazards
  and Vulnerability, which took place in Bangkok, Thailand, in March 2002,
  resulting in the Bangkok Statement: a pledge to promote the protection of
  children's environmental health (see Chapter 1). The Statement urged the
  World Health Organization to support efforts in the region to improve children's environmental health.
- The United Nations Millennium Development Goals published in September 2001 called for a two-thirds decrease in the under-five mortality rate by
  2015, which will require action to reduce illness and death from diarrhoeal
  disease and acute respiratory infections, two leading environment-related
  causes of death worldwide.
- The Budapest Conference held in 2004 focused on "the future of our children" and adopted, through its Declaration, a children's environmental and
  health action plan for Europe (CEHAPE).

15.5 READY ... SET ... GO!

From very small, local, community-based steps to dramatic international accords, children's environmental health continues to gain momentum, expand its audience and increase in significance. There is growing recognition that environmental health is both a right of children and the basis for sustainable development. Simple actions can improve the lives of children and give them the best possible opportunities.

With the goals of increasing public awareness, defining the roles and responsibilities of health professionals, and achieving government buy-in and policy change, four action areas have been defined: communication, education, advocacy and research. Efforts to inform people about children's environmental health have tremendous potential. As people become more informed, health professionals will need more knowledge to answer their questions. As health professional training changes to incorporate the recognition and management of environmental exposures and the particular vulnerabilities of children, health workers will add to the awareness and information sources of parents and of children themselves. As the competence of health professionals increases, they will begin to identify gaps in knowledge and research needs, make recommendations to policy-makers, and advocate for change to protect health and prevent disease. As people become more informed and health professionals more vocal, government officials will set policies that protect children from environmental harm. As governments champion country efforts, national movements will start that will serve to raise public awareness and improve professional education. Action targeted to any of these areas will result in positive effects all around.

15.5.1 COMMUNICATION AND PUBLIC AWARENESS

Communication and public awareness efforts involve a broadly based approach to inform people of all ages and functions, from children to heads of state, leading to an increased understanding of the importance of protecting children from environmental harm. Internet resources are powerful mechanisms to facilitate information exchange, allowing participation of individuals, communities and national groups in global efforts. Both formal and informal actions to raise public awareness have proven successful.

15.5.2 EDUCATION

Actions in the area of education aim to increase the competence of health professionals, especially those dealing with children. They need to learn how to recognize and manage the health effects of environmental exposures, and to break the cycle of exposure, illness, treatment and re-exposure. Physicians, nurses, midwives and other health professionals are in the front line of children's environmental health, and can use their clinical experience, scientific expertise and research efforts to work closely with children of all ages, their children's environmental health can have significant influence, leading to greater public understanding and awareness, improved diagnosis and treatment of environmentally related diseases, and extended advocacy efforts to promote policies that protect environmental health. The general outline of courses offered by WHO is presented below.

15.5.2.1 CHILDREN'S ENVIRONMENTAL HEALTH FOR HEALTH CARE PROVIDER

Contents of a training course
1. New knowledge on the vulnerability of children to environmental hazards
    a) Why are children more vulnerable than adults?
    b) The developing child and the effects of neurotoxicants (lead, mercury, manganese, PCBs)
    c) Lung development and the effects of environmental pollutants
    d) Vulnerability to pesticides: new data and growing concern
    e) Genes as a target for environmental toxicants, malnutrition, micronutrients and toxic effects (including methylmercury, arsenic)

f) The effects of UV radiation on eyes, skin and the immune system

g) Other examples.

2. How, when and where does exposure occur? Environmental threats in specific settings and circumstances, in utero and during childhood and adolescence: "children growing in an adult-size world"

a) The poor home: particular risks (shanty towns); living near waste sites; polluted urban areas; rural areas; street children; parental exposure

b) Where the child plays: playgrounds (outdoors, indoors); recreational areas; hobbies

c) Where the child learns: child care centres; schools

d) Where the child works: cottage industry; factory; rural areas; street vendors; domestic workers; scavengers

e) Where the child is especially stressed: extreme and adverse climatic conditions (e.g. mountains, hot and cold weather); environmental and technological disasters (floods and droughts); war; conflict and postconflict circumstances; refugee camps

f) Exposure of parents: transgenerational effects.

3. Understanding the main environmental threats and setting the priorities for action

a) Access to safe drinking-water and sanitation

b) Indoor air pollution: open fireplaces indoors, environmental tobacco smoke (parents); solvents; moulds; pet dander; other

c) Ambient air pollution and the health of children from rural and urban areas: sulfur dioxide (SO_2); nitrogen oxides (NO_x); diesel fumes; fine particulate matter; lead; benzene; open burning (waste and other); other

d) Asthma and other respiratory diseases in children: role of the environment

e) Traffic-related paediatric pathology. Giving priority to children in township development planning: "child-size traffic." Rural traffic accidents

f) Non-intentional, intentional and environmental toxic exposures

g) Exposure to pesticides: acute and chronic effects

h) Endocrine disrupters

i) Drugs of abuse

j) The working child

k) Lifestyle changes influencing housing, transport and children's social surroundings

4. Assessing the global burden of environmental threats to the health of children

a) The concept of global environmental burden of disease (GEBD) in children

b) Harmonized procedures, tools and methodologies; guidance for assessing the GEBD in children; indicators of children's environmental health

c) Information available in developing and industrialized countries/regions; national profiles

d) Priorities identified (incl. main controversial issues)

5. Controversial issues, dilemmas and knowledge gaps in the area of children's environmental health (CEH)

a) The risks of living near hazardous waste sites, landfills and open burnings

b) Asthma: the contribution of indoor and outdoor environments

c) The potential effects of climate change (emerging infectious diseases and climate refugees)

d) Noise, hearing loss and other health effects in children

e) What is known about endocrine disrupters and CEH?

f) Cancer and environmental factors: how much do we know?

g) Birth defects, reproductive disorders and environmental factors

h) Is there a "safe" blood lead and mercury level in children?

i) The potential effects of exposure to low chronic radiation levels and electromagnetic fields

j) Problems posed by cyanobacteria in water and other contaminants

k) Parental exposure

6. Ensuring the appropriate risk assessment in developing children
    a) Setting environmental guidelines and standards.
    b) Considering variability in exposure and response
    c) Critical windows of exposure
    d) Special consideration of developmental effects
    e) Cumulative toxicity/mixtures, multiple exposures
    f) Recommendations for improved methodologies for exposure assessment and determining health effects

7. Incorporating CEH issues in the work of child health professionals
    a) Recognizing the links between paediatric morbidity and environmental threats in the micro- and macroenvironments of children
    b) Clinical observations: harmonized case data collection and analysis
    c) Taking the paediatric environmental history: from symptoms to etiology to prevention
    d) Detecting emerging diseases and signals of environmental illness in the community
    e) Reporting and publishing observations
    f) Undertaking research studies
    g) Evidence-based interventions: illustrative cases
    h) Communicating with parents, teachers, the community, media, local authorities and decision-makers

15.5.3 ADVOCACY AND PUBLIC POLICY

These activities aim to improve the state of the environment and target policies towards children's health, so that local, regional and national governments act to improve both the environment and the health of children and those around them. In many countries, governments lead the effort. In other countries, policy-makers react to public and professional demands. All levels of society can advocate for children's environmental health and influence policy agendas. A global effort to develop children's environmental health indicators is under way, coordinated by the World Health

Organization and the United Nations Children's Fund (UNICEF). Indicators of children's environmental health offer a tool to policy-makers for determining priorities and measuring progress towards set goals. Governments have the opportunity to join the global initiative on children's environmental health indicators by contacting WHO or UNICEF.

15.5.4 RESEARCH

Promotion of collaborative research in children's environmental health in developing and developed countries is essential if problems are to be addressed in their national and global contexts. The results of appropriate studies can be used in strategies for prevention, intervention, and remedial action, and as a foundation for evidence-based public health policies in countries. Collaborative activities would also result in technology transfer and capacity building, and in the development of a network of trained scientific collaborators throughout the developing world.

15.6 NATIONAL PROFILES AND INDICATORS

WHO has developed indicators of children's health and the environment and other tools to assist countries in assessing the status of children's health and determining the readiness of governments to effect change. A format for doing rapid assessments that may help countries to prepare their national strategies is shown below.

15.6.1 OUTLINE FOR PREPARING NATIONAL PROFILES ON THE STATUS OF CHILDREN'S ENVIRONMENTAL HEALTH

NOTE: please use the headings as subheadings in the country/local profile you develop. Use the questions proposed as a guide for obtaining and collating information and developing an overall assessment of each area. These questions are intended to provide some orientation on the type of information that is relevant for assessing the status of children's health and the environment. Develop up to three paragraphs for each of the under-lined headings, expanding even beyond the questions provided, as deemed necessary. Please take into account for each question the potential gender, rural/urban, cultural and ethnicity issues. Tables necessary to make a point can be annexed. The profiles should cover both existing situations, obser-vations and ongoing activities as well as potential opportunities for actions that could be implemented at the country level. Profiles should be dated: once the initial profile is done, successive ones may be prepared on an annual basis to assess progress made and/or changes observed concerning the status of children's health and the environment in the country.

15.6.1.1 INTRODUCTION

Overview of children's environmental health in the country

Provide a general synopsis of the country's views and position on chil-dren's environmental health, for example, the awareness level of govern-ment officials (especially in the health and environment sectors) and the acceptance of this as a distinct issue.

Key environmental issues

WHO lists the following key environmental risks for children: unsafe wa-ter, air pollution (indoor/outdoor), poor food hygiene, poor sanitation and inadequate waste disposal, vectorborne diseases and exposure to chemi-cals (agrochemicals, industrial chemicals, persistent toxic substances, nat-ural toxins and other). In addition, children's health is endangered by other environmental risk factors, such as: poor housing, environmental degrada-tion and the the so-called "emerging" threats (e.g. global climate change, ozone depletion, radiation, exposure to endocrine-disrupting compounds, and others). Prioritize these for your country according to the impact they have on children's health, development and well-being. Add areas of focus if necessary. Propose a prioritized list of environmental concerns for chil-dren's health in your country.

Key causes of infant and under-tive mortality/morbidity

This information is normally readily available from WHO websites or in the WHO representations in the country. List the top five causes of illness and death for children under one, for children five and under, for children up to 14 and for children as a whole. As the age groups of children vary somewhat from country to country, please define the age group that you are reporting (e.g. some use 18 and under, some 20 and under).

Burden of disease related to environment in children

WHO has some information available on its website (www.who.intlphe/ health-topics, search for "environmental burden of disease") and at the WHO representation. WHO reports that environmental threats may cause up to one-third of the global burden of disease. What does the country report? Are there any significant differences between boys and girls or between rural and urban children? Has this issue been addressed at the country level or does it remain to be done?

15.6.1.2 ECONOMIC STATUS AND ETHNIC GROUPS

Economic spread between poorest and wealthiest

What is the percentage share of income or consumption for the wealthiest 10% of the population? What is the percentage share of income or consumption for the poorest 10% of the population?

Information on high risks/Vulnerable groups and demographic profile of countries

Provide the approximate numbers or percentages of each ethnic population group in your country and the geographic areas they occupy. To what extent are environment and health statistics or any other statistics routinely desegregated by socioeconomic status or ethnicity? Do national environmental or other sectoral policies make specific reference to ethnic groups or to groups that are geographically isolated? Is there any evidence of the impact of ethnicity or socioeconomic status on the burden of disease related to environmental threats? Are there any activities on ethnic minorities (and their children) undertaken by international institutions or nongovernmental organizations to which an environment and health component might be added?

15.6.1.3 NATIONAL GOVERNMENT ROLE

National policies

Are there specific national policies or stated priorities that support the protection of children's environmental health? Are there specific national policies or stated priorities that seem to run counter to the objectives of increasing protection of children from environmental threats (e.g. lax pesticide or toxic chemical regulations, persistence of lead in gasoline despite the proven health benefits of removing it)?

Health sector

How does the health sector address environmental health in general and children's environmental health specifically? Is there legislation to protect public health from environmental hazards and is this legislation well-implemented? Is there any action to protect vulnerable sub-populations or children in particular? Are the medical, nursing and health-care professional communities informed and/or trained on environmental threats to human and-more specifically-on children's health? Are there health facilities that promote environmental health or children's environmental health? Describe the differences in approaches to environmental health in rural and urban settings. In the specific area of chemicals, is there a Poisons Centre in your country or a toxicology or other unit that deals with toxic exposures in children? Where are poisoned children seen and treated? Are chronic, low-level exposures to chemicals in children being considered? Has any action been taken concerning the potential effects of persistent toxic substances (and POPs)?

Environment sector

Discuss the country's environmental legislation and its level of enforcement. Is human health considered by the environmental legislation and/or is protecting human health part of the mandate of the environment ministry? Are there any specific considerations concerning children? Are specific media, such as water, air, soil, food, or chemical safety covered by environmental legislation? If so, list which media are covered and list any gaps. Does the environment ministry coordinate well with other ministries, such as health or education and, if so, which ones? Has the country signed the international conventions/treaties dealing with toxic chemicals/pollutants (e.g. The Stockholm Convention on Persistent Organic Pollution, The Basel Convention on the Control of Transboundary Movements of Hazardous Wastes and their Disposal, and The Rotterdam Convention for Prior Informed Consent)? Have the actions taken in the context of these conventions/treaties considered the potential impact on children's environmental health?

Education

What is the level of literacy in the country? How many children go to school, and up to what levels? Is attendance in the schools required up to a certain age? Are there differences in male/female school attendance? For elementary school and for high school, what are the opportunities for health and environmental education? Is there an environmental or a health curriculum taught in these grade levels? If so, are these taught in both rural and urban schools? Would environmental health education through elementary schools be possible and/or acceptable in the school systems?

Other pertinent ministry/sector

If applicable, list other pertinent ministries or governmental agencies that deal with children's health and the environment (e.g. in certain countries some of the environmental issues may be regulated through the ministries of agriculture, industry, youth, social well-being or others). In many countries there are ministries of culture, science, education, welfare, and family and youth that may play a role in the protection of children's environmental health. What are the ministries or agencies at the national government level which would playa role in implementing a national action plan on children's environmental health? List and describe the role they play.

15.6.1.4 SOCIETY ROLE

Communities

Do the governmental units at the community level (e.g. county seats, communal or city governments) playa role in the protection of environmental human health—and more specifically—children's environmental health? If not, what role could they play, or might they take, at a local level to better protect children from environmental threats? Do they have the ability to pass local legislation? Are they charged with enforcing national legislation? Could they be encouraged to carry out public information campaigns on children's environmental health?

Nongovernmental organizations (NGOs)

Do NGOs playa strong role in building stakeholder input and public participation? What are the key NGOs (both national and international) involved in activities aiming at the protection of children's environmental health, organizing national campaigns on children's environmental health or promoting children's chemical safety? If none has been doing this, which one could eventually be interested in this area? What roles might they play?

Professional associations

Do professional associations play a strong role in building stakeholder input and public participation? What are the key professional associations (both national and international) that would become involved in children's environmental health? (e.g. paediatric, medical, toxicological, family doctors, occupational medicine, nursing, primary health care, and any other societies). What roles might they play?

Academia

What academic institutions (e.g. academies, post-graduate schools) could promote children's environmental health through research, advocacy, publications, medical education (of medical and postgraduate students and continuing medical education), development/use of children's environmental history taking, and development/use of indicators? What role would each play?

Private sector

Are there any private companies that would likely be interested in promoting the safety and health of children in the country? For example, pharmaceutical, hygiene and cosmetic products companies, agricultural chemical companies, water companies, food and beverage producers? What roles could the private sector take-considering after all ethical aspects involved-in the different areas (e.g. financing activities, public advertisements, educational campaigns, or advocating in favour of national legislation)?

15.6.1.5 SCIENCE

State of the science in the country (in relation to children's environmental health)

Has anyone in the country conducted research and published results on topics related to environmental health or children's environmental health (e.g. on the risk factors mentioned above, on children's settings, on specific topics such as chemical safety and poisonings)? Name the country's science ministry or unit in the government that conducts research and publishes findings. Is environment or health legislation based on scientific findings?

Capabilities to conduct research

What institutions promote science and research in the country? Does the national government invest in research and development? What type of scientific publications are released in the country? Is financing available to support research at universities, hospitals, laboratories or other facilities? Which institutions would most likely be interested in research on children's environmental health?

Research needs

List the top priority research needs around the topic of children's environmental health in the country. Is research on these topics under way? Are there barriers to conduct this research and, if so, what would help overcome the barriers? What are the needs? What are the top three ways in which an international organization or other countries or organizations could support research?

15.6.1.6 DATA AND REPORTING

Information systems and centres

Does the country have a centralized information gathering function on children's health data (e.g. health surveillance system, clinical case recording)? Does the country have national or private information centres, for example on health, demographics or environment? Does the country require reporting of certain paediatric diseases to support public health surveillance and disease prevention and, if so, how is that information gathered and where? Are there poison control centres in the country and, if so, do they record incoming and outgoing information in a harmonized manner? Does the country report indicators on environment or health? Does the country put out regular reports on disease, public health or environmental conditions? If so, how are they accessed by the public?

Data quality

The WHO national offices are most likely involved in data gathering on health, and local UNICEF and UNDP offices probably work on information collection systems, as well. Do these offices judge data quality as good enough to be useful and representative? Are there other entities that collect data on health, environment or status of children in the country? Can the national work on Millennium Development Goals help to clarify and address barriers to data quality in the country?

15.6.1.7 COMMUNICATION

Avenues of communication

What are the most effective means for disseminating information in the country (e.g. television, radio, newspaper, and role-playing)? Are these the same for both rural and urban settings? If not, list by rural and urban. What are the most effective means for communications through schools, adult literacy programmes, country or local governments? Are there other innovative means of communication, for example through local libraries, street theatres, radio/TV educational "soap operas," fairs or other local events?

Success stories in communication

Do you know of any local success stories in widespread communication on important topics related to health and the environment? (e.g. use of radio-based literacy programmes targeting children in rural areas may increase adult and child literacy and lead to a decrease in child agricultural workers and improve matriculation in rural schools). Could these successes be repeated, this time carrying a message for children's environmental health?

15.6.1.8 CONCLUSION

Summary of the country status of children's environmental health and opportunities for action

Given your findings, in a page or less, summarize your assessment of the country's potential, capacity and interest to take action to improve the environmental health of its children. What specific actions in this area are recommended? What are the areas/issues for natural success? What are the areas/issues where urgent actions are required? What are the key barriers or areas that need to be addressed to achieve success? Who (individuals and organizations) are the key players?

Annexes

Please provide any samples of useful or illustrative materials, such as educational, awareness building, information gathering, data collection forms, educational programmes, photographs, maps, charts, other.

15.7 A SPECIAL INVITATION FOR NONGOVERNMENTAL ORGANIZATIONS

Nongovernmental organizations (NGOs) have excellent opportunities to promote awareness of environmental hazards to children. They can use radio, television, newspaper advertisements, street theatre, health fairs, and other innovative ways to improve public awareness, increase training for health care professionals, enhance access to information and advocate better policies. The information era affords enormous potential to broadcast messages, reaching urban and rural areas alike. Religious leaders can also have a strong impact, reaching receptive audiences who want to take actions that benefit children.

Many NGOs actively promote activities that support the environment, from nature conservation to sustainable development, but there are relatively few in the field of environmental health or environmental medicine, particularly dealing with the special vulnerability of children. Two organizations that have reacted to the realization that children are particularly susceptible to environmental risks are the International Society of Doctors for the Environment (ISDE; www.isde.org) and the International Network for Children's Health and Environmental Safety (INCHES; www.inchesnetwork.net). Some countries have national networks for children's environmental health, such as the United States Children's Environmental Health Network (CEHN; www.cehn.org), the Canadian Institute for Child Health (CICH; www.cich.ca) and the German Network for Children's Health and Environment (www.kinder-agenda.de). The web sites of these national networks are rich in information resources and links.

15.8 CONCLUSION

Pollution and other environmental threats do not recognize borders. Action is required at all levels: even local, community-based activities may end up having great influence around the world. The history of children's environmental health demonstrates how local actions may have a global impact: a nongovernmental organization with a clear mission convinces a minister of environment, who motivates an international declaration, boosting children's environmental health into the mainstream international agenda. Not only do actions at the different levels affect those in the immediate area, they also create energy for public good with worldwide benefits. Everyone at every level can do something to improve children's environmental health and advance sustainable development while contributing to the health, increased productivity and wellbeing of children around the globe.

REFERENCES

1. Tips to protect children from environmental risks. US Environmental Protection Agency, 2004 (http:j jyosemite.epa.gov jochpjochpweb.nsfjcontentjtips).
2. Healthy environments for children alliance. Geneva, World Health Organization, 2003 (http://www.who.int/heca/enf).
3. Children's environmental health. Geneva, World Health Organization, 2004 (http://www.who.int/ceh/en).
4. Commission for Environmental Cooperation of North America. Cooperative agenda for children's health and the environment in North America. Montreal, 2002 (http:j jwww.cec.orgjfilesjpdfjPOLLUTANTSjChildren_coop_agenda-en.pdf).
5. A worldfit for children. New York, United Nations, 2002 (http://www.unicef.org/specialsess ion jwffcji ndex.html).

AUTHOR NOTES

CHAPTER 4

Funding
The CAMP Genetics Ancillary Study is supported by U01 HL075419, U01 HL65899, P01 HL083069 R01 HL 086601, from the National Heart, Lung and Blood Institute, National Institutes of Health. The methylation data described here was funded by 5RC2HL101543-02 and NIEHS grants 5P30ES007048 and 1K01ES017801. Also supported in part by the Division of Intramural Research, National Institute of Environmental Health Sciences, National Institutes of Health, Dept of Health and Human Services, US. The Norwegian Mother and Child Cohort Study is supported by the Norwegian Ministry of Health and the Ministry of Education and Research, NIH/NIEHS (contract no NO-ES-75558), NIH/NINDS (grant no.1 U01 NS 047537-01), and the Norwegian Research Council/FUGE (grant no. 151918/S10). KCB was supported in part by the Mary Beryl Patch Turnbull Scholar Program. The funders had no role in study design, data collection and analysis, decision to publish, or preparation of the manuscript.

Competing Interests
The authors have declared that no competing interests exist.

Acknowledgments
We thank all CAMP subjects for their ongoing participation in this study. We acknowledge the CAMP investigators and research team, supported by NHLBI, for collection of CAMP Genetic Ancillary Study data. All work on data collected from the CAMP Genetic Ancillary Study was conducted at the Channing Laboratory of the Brigham and Women's Hospital under appropriate CAMP policies and human subject's protections.

Author Contributions

Conceived and designed the experiments: CVB FG BR. Performed the experiments: WN SEH FG BR. Analyzed the data: CVB KDS BRJ XW WQ SL FG BR. Contributed reagents/materials/analysis tools: VC WN SEH CO DN KCB FM AL RL RS SW SL FG BR. Wrote the paper: CVB KDS FG SL BR.

CHAPTER 5

Funding

This study was funded by the Netherlands Organisation for Health Research and Development (ZonMw: project no. 22000128). L. Duijts received funding by means of a European Respiratory Society/Marie Curie Joint Research Fellowship (MC 1226–2009) under grant agreement RESPIRE, PCOFUND-GA-2008-229571. V.W. Jaddoe received additional grants from the Netherlands Organization for Health Research and Development (ZonMw - VIDI). The funders had no role in study design, data collection and analysis, decision to publish, or preparation of the manuscript.

Competing Interests

The authors have declared that no competing interests exist.

Acknowledgments

The Generation R Study is conducted by the Erasmus Medical Centre in close collaboration with the School of Law and Faculty of Socioeconomic Sciences of the Erasmus University Rotterdam, the Municipal Health Service Rotterdam area, Rotterdam, the Rotterdam Homecare Foundation, Rotterdam and the Stichting Trombosedienst & Artsenlaboratorium Rijnmond (STAR-MDC), Rotterdam. We gratefully acknowledge the contribution of children and parents, general practitioners, hospitals, midwives and pharmacies in Rotterdam. We gratefully acknowledge the contribution of all Generation R participants and the professionals at the participating well-child centres in Rotterdam, especially Inge Moorman, MD.

Author Contributions
Conceived and designed the experiments: ADM JCW HJK JCJ HR. Performed the experiments: EH ADM. Analyzed the data: EH RJPV. Wrote the paper: EH RJPV ADM JCW LD VWJ AH HJK JCJ HR.

CHAPTER 8

Acknowledgments
We thank William De Vore, Dursun Peksen, and Richard Dalton for their review. We acknowledge the assistance of Betsy Shockley, Supervisor of Childhood Lead Poisoning Prevention Program, Memphis and Shelby County Health Department.

Author Contributions
Cem Akkus reviewed the articles and drafted the manuscript. Esra Ozdenerol worked with Cem Akkus in the preparation and editing of the manuscript.

Conflicts of Interest
The authors declare no conflict of interest.

CHAPTER 9

Author Contributions
The manuscript was written through equal contributions of all authors. All authors have given approval to the final version of the manuscript.
The authors declare no competing financial interest.

Acknowledgments
We thank the Michigan Department of Community Health, Childhood Lead Poisoning Prevention Project for providing the blood lead data used in this study. We thank the Robert Wood Johnson Foundation Health & Society Scholars program for its financial support

CHAPTER 10

Acknowledgments

Thank you to the CHARGE participants for helping make this research possible. This work was supported by grants from: NIEHS R01 ES015359, NIEHS P01 ES11269, EPA STAR #R829388 and R833292, The UC Davis Division of Graduate Studies and the UC Davis MIND Institute.

Competing Financial Interests

The authors have no competing financial interests.

CHAPTER 11

This study was supported by the National Institute of Environmental Health Sciences (grants 5P01ES09600, P50ES015905, and 5R01ES08977), the U.S. Environmental Protection Agency (grants R827027, 8260901, and RR00645), the Educational Foundation of America, the John and Wendy Neu Family Foundation, the New York Community Trust, and the Trustees of the Blanchette Hooker Rockefeller Fund.

The authors declare they have no actual or potential competing financial interests.

We are grateful to the families of northern Manhattan who have so generously contributed their time and effort to the study.

CHAPTER 12

This research was supported by the September 11th Fund of the New York Community Trust and United Way of New York City; the New York Times 9/11 Neediest Fund; the National Philanthropic Trust; National Institute of Environmental Health Sciences grants ES09089, 5P01 ES09600, and 5R01 ES08977, and U.S. Environmental Protection Agency grant R827027.

The authors declare they have no actual or potential competing financial interests.

We thank C. Dodson, W-J Wang, K. Lester, and L. Stricke.

CHAPTER 13

Competing Interests
The authors declare that they have no competing interests.

Author Contributions
EDH, GS, KC, LB, MK, MV, NVL, TN, and WB designed the study. EDH, GS, IS, NVL, and VN did the field work. AC and AD analyzed BFR levels. LB, MK, and TN performed statistical analysis. MK drafted the manuscript. All authors critically revised the manuscript.

Acknowledgments
The study was commissioned and financed by the Ministry of the Flemish Community (Brussels, Belgium) and FWO grants: G.0.873.11.N.10 / 1.5.234.11.N.00. Michał Kiciński has a Ph.D. and Isabelle Sioen a postdoctoral fellowship of the Research Foundation – Flanders (FWO).

INDEX